THE WAY
OF THE GUN

THE WAY
OF THE GUN

A Bloody Journey into the World of Firearms

IAIN OVERTON

HARPER

NEW YORK • LONDON • TORONTO • SYDNEY

A hardcover of this book was published in 2016 by
HarperCollins Publishers.

Originally published in a different form as *Gun Baby Gun*
in Great Britain in 2015 by Canongate Books, Ltd.

FIRST HARPER PAPERBACK EDITION PUBLISHED 2017.

All images © Iain Overton

Library of Congress Cataloging-in-Publication Data has been applied for.

ISBN 978-0-06-234607-0 (pbk.)

17 18 19 20 21   OFF/LSC   10 9 8 7 6 5 4 3 2 1

For G.
For S.
For A.

In memory of my grandparents
Carolina Bernal (1917–2014)
Pablo Antonio Bernal (1919–2014)

CONTENTS

PREFACE

A few years ago I created an email alert from Google News. It was a simple, three-letter word – 'gun' – and in the weeks and months that followed, each and every morning an email would ping through. Each offered up a dozen news stories from the previous day – all on the matter of guns. And yet, despite the email being one that should cover stories about guns around the world, each day there was one constant.

All the stories were about the United States.

This Google alert's focus on guns in America was, to some degree, disconcerting to me. As a journalist and filmmaker I had travelled to dozens of countries around the world where guns, and the violence they bring, were a daily occurrence. From Central and South America to Africa and the Middle East, I've seen countless places where the gun wreaks havoc and claims untold lives. These are often places riven with corruption, lands ruled by despots, countries marked by a failed rule of law and the ugly stamp of poverty. These are places where lives seem, all too often, to be as cheap as a bullet.

My Google News feed should have picked up on these dark realities. But it is gun violence in the United States, and that country alone, that dominates. This is partly because the world's media dances to a tune all too often orchestrated by that global superpower. But it is also because the US has, quite simply, the most unique relationship with guns of any other country I know.

For a start, it is by far the most dangerous place for gun violence in the world of developed economies. There are over three times as many Americans killed by guns per capita than in any other wealthy country, and over ten times the rate in comparable countries such as Japan, France and Britain.

That is not all. It is, globally, the country with the most gun suicides. It is a land with the highest rate of gun ownership; a nation with the highest number of gun imports and exports per year. It's a land where, since the terrible shootings of Sandy Hook in December 2012, bills have been introduced that would legalise the carrying of concealed weapons in bars, allow teachers to carry guns in schools, and even make businesses liable for customers' injuries or deaths if they did not permit their clients to enter the shop carrying a loaded weapon – the only country in the world where laws governing the carrying of firearms have been relaxed, not tightened, after a gun massacre.

The US, quite simply, *is* the land of the gun. It's the Gunfight at the OK Corral. It's Harry Callahan's 'Go ahead, make my day'. It's the sniper bullet that killed JFK. It's Colt and Smith & Wesson. Custer's Last Stand and *Annie Get Your Gun*. Quentin Tarantino, drive-by shootings and SWAT teams. It's the deaths of John Lennon and Martin Luther King, and the oiled-up swagger of Stallone and Schwarzenegger. It's Sandy Hook, Columbine, Virginia Tech, Fort Hood, Aurora and school shootings in Oregon. And it's endless, endless figures about guns and violence that are so disturbing they just blunt you.

But perhaps it is best to start with those hard facts. Just take one decade of gun statistics in the US.[1]

Between 2004 and 2013, according to data from the US Bureau of Justice, over 4.5 million people were the victims of gun crime in the United States. Of these, the FBI reports almost 95,000 firearm homicides. It's a figure that is over five times the number of knife murders.

The FBI data also shows that, between 2004 and 2013, 66 children under the age of one were shot and killed in the US. A further 363 children between the ages of one and four died from a gunshot wound. In total, in that decade 4,207 American children sixteen or younger were killed by guns.

This is not all. According to data from the Centers for Disease

Control and Prevention, their upper figures show that just under a million Americans may have been shot but survived (though this figure should be taken with a pinch of salt – the truth is that no one knows how many people are shot and injured every year in the US). If you are American, then I am sure you know of someone who has been shot. If you are European, the chances are you don't.

There are other figures too: the terrible and personal moments of gun horror. In that decade over 185,000 Americans shot and killed themselves – deaths that never made the global news.

Such violent death and injury has been fuelled, without question, by America's love for the gun. In that decade the US imported over $10 billion of arms and ammunition and exported about $5 billion worth, and the Department of Defense spent nearly $11.5 billion on guns and ammunition.

What does this add up to? Well, it leads to one thing, at least. You can't even begin to write a book about guns without writing a book about the US. And it means that any book that charts a global journey into the communities that proliferate around the gun has to acknowledge that the world's greatest superpower has embedded within its constitution the right to bear arms.

So it was that, writing this book, I was to meet many American gun owners. And many of them, even with the best intentions, would cock their heads to the side and say, in as polite a way as they could: 'You see, our approach to guns is different from Europe's.' It seemed to many of them that I lost my right to write about guns simply because I was not an American.

But, to me, it was better that I was not an American writing this book. My view could remain, as much as possible, uncontaminated by the fervour and prejudices that exist – on both political sides – of the US gun debate. Perhaps more important, it also enabled me to get the wider view. As a journalist and a filmmaker, a weapons researcher and a former gun-club president, a hunter and a sports shooter, it enabled me to put America's love for guns in a global context. To see what impact the Second Amendment had on inter-national smuggling networks, arms trade deals, UN gun conventions. And, perhaps most important for me, it gave me the chance to find a wider meaning in the countless Americans who have died at the end of the gun.

In essence, this is what this book is about. It is an attempt to understand the world of the gun, without getting caught up in the endless debate in America about the right to bear arms. It is a journey navigated through facts and realities, not politics and prejudice. A passage undertaken open-minded and yet, simultaneously, deeply disturbed by what I had seen.

It is, quite simply, the way of the gun.

I. INTRODUCTION

1. THE GUN

Brazil – a murder in São Paulo – a child's sorrow and a dead mother – the descent into a police arms cache – a revelation – a journey conceived – Leeds, UK – a secret museum and a meeting with an expert – to a Swiss canton to visit an oracle

It began with a death.

The five-year-old had lain alone with his lifeless mother all night long, curled up at her cold feet. It was only when the thin light of dawn lifted some of the darkness from the bedroom that the neighbours had heard the boy's cries. And only then did people realise what had happened in those sunless hours before.

The bullet had entered the left side of the young woman's temple and exited at the back of her head, splattering flecks on the leprous wall. There had often been wild-voiced arguments in that cramped house, but no one ever thought it would come to this.

After the boy was found the police arrived quickly, but the murderous lover had already fled that Brazilian city and, like the gun he had used, he was nowhere to be found.

By the time we reached the quiet roadside home the child had also been spirited away, covered in a blanket – lifted from his dark *pietà* and carried out into the light. His mother was still inside.

Cars passed, leaving São Paulo for the north, and we stood awkwardly and watched them go. They slowed down and watched us too, a huddle of cops and a documentary crew crowded beside a white ambulance that was never really needed. A dog barked in the distance, and I took out my video camera and walked inside.

The dead woman had run a small shop out the front, and it was filled with packets of coloured sweets and warm bottles of luminescent drinks. On the counter was a tray of Catholic pendants, which she had sold to the weary lorry drivers who would stop here. But these plastic icons had not helped her last night, and now she lay beyond, past a dusty glass counter, down a narrow corridor, there in a pool of silence.

They say death smells sweet. That's what I thought as I walked into her bedroom. A taste touched my mouth and reminded me of the orange-tinted bottles that lined the shop's walls or the citrus chocolate puffs that lay neatly arranged in their shiny little packages. The air was thick with this smell. It had been over twelve hours since she had died, and this was the start of summer.

Her name was Lucicleide, and she was naked. I was not expecting that, but death rarely grants us dignity, so her breasts hung to the side and the rest was uncovered. There was not much blood, save for a smear above her pinched, sallow face. Finding a corner, I set up my tripod and got to work; the police did not tell me to stop filming, but by now I was not even sure why I was doing this. My footage would never end up on the evening bulletins. The film I was making with Ramita Navai – an Anglo-Iranian journalist who was used to witnessing such things – was about the toll of violence in one of Brazil's deadliest cities, but Britain's *Channel 4 News* could never show such intimate and murderous detail.

I felt I had to do something, though.

So I focused on her unfurled hands and on the trinkets that lined the top of her chipped cabinet and shifted the lens onto the face of a purple bear I imagined her lover had once bought her. And the whirr of the tape in the camera took the edge off the awkward quiet of the room. I carried on filming until the forensic examiners wrapped her in a heavy blanket, and all I could think of as she was lifted heavily up, covered like her son had been, was how hot it was to have such a blanket to sleep under.

We followed the body out into the light and slipped back into our car. Then, after waiting for the coroner's van to slide away, we too pulled out and drove south, following an unhurried squad car back to the city's police headquarters. And none of us spoke.

The building was low-slung and squat, built in the way Brazilian architects love: concrete, slats and shutters. Municipal chic that was made markedly less chic and more threatening by the armed police standing behind its long glass front. The steps leading up to it were broad and shallow; they twisted in an arc and made the walk to the doors of justice a slow one.

An image of São Paulo as a dystopian city, something out of a *Judge Dredd* comic, came to mind. This was its rigid heart of order and legal retribution, but policemen stood, their arms cradling dull metallic weapons. The reason for this was clear. In one year alone there were over a thousand gun murders in this city, in waves of crime so violent they had caused schools to close and municipal bus routes to change.[1]

It was small wonder that this governmental building was so foreboding. Dozens of police officers had been killed, caught up in the endless drug wars that blighted this land. The guards at the entrance were taking no chances. They were heavily armed – police assault rifles slung across their riot shoulder pads and bulletproof vests lying underneath.

Passing through scanners and scrutiny, we emerged on the other side. There we were met by Colonel Luiz de Castro. A short man with dark, tightly cut hair, a firm jaw and a precisely ironed shirt, he looked like someone born to be in the service of the state. Greeting us with an iron handshake, he was quick to address why we had come here: to see São Paulo's seized-gun repository.

'A few years ago we had a gun amnesty,' he said as if delivering an order, his voice a staccato drumbeat. 'Here we have about 20,000 weapons confiscated or handed in. We offer between $50 and $100 for each one.'

The colonel spun on a heel and led us at pace down a long corridor, lit by naked and glaring fluorescent strips. The sandpaper walls here were bare and the floor scuffed, and as we descended slowly into the sodium-coloured belly of this bureaucratic beast, the colonel walked ahead, his boots sounding the mark of his passage. Then he stopped at a grey door and motioned us inside. Beyond lay a small room with a few computer terminals, in front of which sat uniformed officers. They were inputting data and looked up at us with the eyes of people whose lives were spent in rooms without sunlight. Across from them lay a caged door.

The colonel called out, and a shadowed face appeared; keys were turned, and the door swung outwards. We walked into the semi-darkness.

Beyond lay thousands of guns. Every surface was filled with them – the walls lined with wooden, narrow boxes, like a mail-sorting office, each pigeonhole containing a gun with a small paper label attached. Space here had run out long ago, and the guns spilled out onto the counters and the wooden chairs that spread across the floor. A door led on to another room and then another, and the scene was the same in each.

There were semi-automatics from North America; hunting rifles from China; a 9mm pistol from Germany; an old blunderbuss from England. There were home-made handguns and high-tech machine-guns. Black guns so corroded with time you imagined them wielded by slave owners in long-shut-down plantations. There were even some with *Polícia* stamped on them, because when Brazilian gangs kill a policeman the prize is that downed cop's sidearm as well as his life.[2]

Then it struck me how, like the clichéd six degrees of separation, this graveyard of guns was somehow more significant than just what was visible in this narrow space. Each gun here, either through maker or victim, shooter or seller, was somehow linked to a bigger story – each connected to the outside world in a deeper, more nebulous way.

Here were revolvers bought with taxpayers' money and ordnance left over from long-forgotten wars. Police pistols and army handguns, sports rifles and hunting shotguns from all over the world, many of them tainted with the stain of murderous deeds. The microcosm of life – of law and protection, violence and vengeance, leisure and provision – was laid out in these shadows.

In a sense this lair of guns was a symbolic image for all of the human rights tragedies I had ever been trying to explain as an investigative journalist and a human rights researcher. And the idea for this book was conceived in that moment – a desire to trace the gun's pathway from its metallic cradle to its blood-tinged grave. A journey to discover the lifecycle of the gun and, in so doing, to understand a little bit more about death and maybe, even, a little about life.

There are almost a billion guns in the world – more than ever before. An estimated twelve billion bullets are produced every year. Over a hundred countries have their own gun industries, and twenty nations recently saw children carrying guns into conflicts. In this new millennium, AK47 rifles have even been sold for as little as $50.[3]

These are hard facts that have harder consequences. And yet, despite how shocking these numbers are to hear, before I began to research the world of the gun, these were facts I did not know. Perhaps this is because the gun remains all too often forgotten in our media and news. Throughout my career I had frequently reported on the harm wrought by firearms, but I had never actually done a report on the gun itself. It was a bit like the face of evil: you knew it was there, but you felt a little foolish mentioning it. With other weapons, it was different. 'A kitchen knife? He was beheaded? My God, that's terrible.' But with a gun it was more like: 'Of course there's a gun.' Guns were just there, remarkable but unremarked upon.

In Lucicleide's shaded bedroom, I had seen one face of the gun: the way it can take a life. In São Paulo's police headquarters, I had seen another: the way the police seek to contain and control guns. But these were isolated images, scattered pieces. Seeing them on their own did not answer questions such as: Who made those guns? How did those police pistols end up being used in a killing? Who profited from the sales of those Uzis?

I knew some of the answers. The assignments and campaigns I had worked on had been diverse enough to allow this. War correspondents usually just witness the harm that guns cause. Arms-trade campaigners often focus on the world of immoral governments. Investigative journalists seek to expose corrupt gun sellers. I had earned a living carrying out all of these roles, and so had seen glimpses of such things and more. Reporting on trafficked women in eastern India or filming the slums of Buenos Aires, seeing the impact of violence in the borderlands of Mexico or recording the tense diplomatic stand-offs between China and Taiwan, I'd seen the gun in many of its varied colours. But there were gaps. I knew little about the world of hunters. I had never met a sniper. Never been inside a gun factory. What I wanted was to bring these parts together – to see the whole, the gun as a sum of its many parts.

Of course, such an undertaking was ambitious. People would suck in their breath when I told them what I wanted to do. Certainly it was global. Too much of the media has fixated on the US's relationship with guns and that alone, but I wanted to take in the wider view. Guns in the US showed, to me, just the tip of a bloody iceberg.

This was, then, a journey born from both memory and new experiences, one where I had to revisit worn notebooks as well as tread the carpets of endless airports on my way to yet another killing. And through doing so I sought to weave a complete tapestry of the impact of guns on our world – where the thread of a moment lived in one city might unexpectedly find itself tied to a visit planned in another, far away.

The view I sought was certainly too big to take in without some sort of plan, so I decided to divide my research into the communities the gun impacted. There were those directly harmed by firearms – the dead, the wounded, the suicidal; those who used guns to exert a form of power – murderers and criminals, police and armed forces; the people who used these weapons for pleasure – hobbyists and hunters; and those who sought to profit from their sale – traders, smugglers, lobbyists and, ultimately, the manufacturers. I planned to approach each community in turn, merging memories and interviews, new trips and research, to grasp fully what it was like to live, and die, under the gun's shadow.

To begin, though, I wanted to understand a little bit more about firearms themselves – to see their historic place in the world, how they evolved and how they have influenced the unfolding of history. So I arranged to travel northwards from my home in London, to the largest museum collection of guns in the world – to the Royal Armouries in the English town of Leeds.

The gun collection at the British National Firearms Centre started almost four hundred years ago. It was originally dreamed up by King Charles I, a hapless monarch who wanted to give some uniformity to his kingdom's procurement of arms. Since then, the centuries have added to the collection; today the armoury boasts the largest number of unique rifles and handguns kept anywhere under one roof. If there was one place to begin a deeper understanding of the world of the gun, this was surely it.

So, on a blustery day in spring, the senior curator there, Mark Murray-Flutter, agreed to meet me at the entrance of the public museum. A large and effusive man, he greeted me in a flurry of great strides and smiles. He held out his left hand to shake me by my right and it confused me; I looked down. Instead of flesh I saw a prosthetic limb. Ex-military, I thought: the price a man pays for being too close to guns. He ignored the look on my face.

Without explaining where we were going, Mark turned and led me away from the municipal grey building at a brisk pace, his tie fluttering. We walked down a wind-filled road under heavy, tea-coloured clouds and there, through an unnamed and unmarked door, we crossed into a windowless space lined with steel and concrete. Beyond was a metal detector and an armed guard asking, through a bulletproof window, if he could see my passport. Then there was a body search. Finally, we entered a cavernous space where the public rarely goes.

'Here you are,' Mark said, a smile widening on his face. 'Where it all is.'

Guns. Thousands of them. They filled the cavernous room like squat metal insects, sleeping before an ugly dawn – hunched, silent and demonic. Under the chrome light you could see row after row of every type of firearm imaginable. There they were, oiled and fierce on the floor. There, neat and polished on racks. Hung on wall brackets, put away on shelves, slid deep into recessed drawers. It was like Borges's infamous library, but here were guns not books – over 14,000 in steel and wood and brass.[4] And here, unlike the police repository in Brazil, the guns were ordered and neat – their potential anarchy contained.

It smelled like history: gun oil and the ghosts of cordite. These weapons spoke of past wars and long-forgotten conflicts, because the curators had tried to get their hands on every type of gun ever produced, within reason. When the British used to mass-produce rifles they would dispatch the prototype – the first edition – to the armouries. There they were stamped with a thick layer of copyright sealing wax and stored away. Elsewhere machines got to work and churned out copies in their millions, and the prototype's offspring wound their way to the foothills of the Himalayas and the steaming jungles of Africa, as this little nation of shopkeepers traded and slaughtered its way into Empire.

The origins of all of that violent shame and bloodied history could be seen here; and this was just the collection of Britain's guns. There were others, too. Here was the United Nations of firearms – it was almost a case of naming a country and a gun from there could be conjured up.

'The best way to think about this place is as a library, but instead of having books you have guns,' Mark said, offering me a cup of tea. A reasonable, softly spoken man, he was not into weapons, he explained. At least not for what they were per se; rather this wounded scholar liked what they represented. He was a social historian, fascinated by how firearms fitted into society. If he had one interest, it was their ornamentation, their decorative appeal. In this way he saw himself as a benign curator – not a man who would view this room in terms of gun control, how many lives taken, how many liberties defended. Rather, he was interested in their meaning.

'I'm fascinated by the use of firearms as a status symbol, as diplomatic gifts, as love tokens,' he said, education in his voice. 'How they can show people you have arrived. Certainly this is true in the world of those who own shotguns – the higher you go up that economic ladder, the less it's about the cost, the more it's about the ostentatiousness of the design. The Russian oligarchs, the Mexican drug gangs who gold-plate their guns, they are trying to show that they are all-powerful.'

We spoke about facts. But, in a way, when it came to Mark giving a broad introduction to guns, there was not that much to say. In this world the devil was in the detail. What calibre, what model, these were finer points that many gun enthusiasts fixate on – but not ones that captured my attention. I couldn't get excited about the small tweaks made to a handgun to sell a newer, deadlier version. I was more interested in what these guns did.

Just as well, really, because when it came to the basic physics of the firearm, Mark said things hadn't really changed since the four-teenth century. All a gun needs, he explained, is a barrel, a missile, a means of projection, a form of ignition and a way to point it. All the developments since these principles were first conceived were pretty much just perfecting this process.

'They may be lighter, more compact, but they are fundamentally the same,' he said, leaning forward over his mug. 'There have been two major step changes in the development of firearms. The develop-ment of the self-contained cartridge in the early nineteenth century and the gun that can fire automatically – developed by the British – by Maxim.' Perhaps this is what lies beneath the enduring popu-larity of guns, I thought. The fact that there's an alluring simplicity to how they work.[5]

Finishing his tea, Mark rose from the table and told me to follow. He handed over a pair of white gloves, and then, like vicious mime artists, we entered the stacks. There he began to pass me rifle after rifle, with a disconcerting casualness.

Closest to us was a Gardner gun – a five-barrelled, hand-operated machine-gun, fed from a vertical magazine. As the crank turned, he explained, a bullet was loaded into the breech, the bolt closed, and the gun fired.[6] Invented in 1874, it was part of a major landmark in the development of the gun. There were even men who saw civilisation's face in this mechanised operation purely because the Gardner gun worked on the principle of serialisation. As such the machine-gun was seen, by some, as a product of a rational culture. By default, cultures that could not create such a killing weapon were deemed less civilised, and so open to imperial rule.

Such men would have been impressed here, because in this fortified chamber the walls were lined with sub-machine-guns. Anti-aircraft guns, first designed to combat the use of observation balloons in the American Civil War, also stood to the far left. To the right there were Chinese DShKs; a gun mounted on wheels, called, affectionately, 'Sweetie'. These stood beside a low line of recoilless rifles once used as tank busters. And there, on the end, were Russian rifles that fired underwater. Civilisation's progress laid out in deadly metal.

These guns all told a story in their own way. They spoke of how rifles and pistols had turned the course of history. How the assassin-ation of Archduke Ferdinand in Sarajevo unleashed the First World War. How the killing of Martin Luther King pushed the US closer to equal race rights. They spoke of how the gun has helped bring advances in industrial production methods and advanced modern medicine. And they all spoke of death.

There, on the far wall, one rack held a familiar shape: the long, curved magazine, the wooden stock, the iron sights. A gun that could fire automatically like a machine-gun, or could let loose single shots, like a sniper rifle; that could be chucked in a river and dragged in the mud and still not jam; a weapon so popular that tens of millions of them have been made. It was the Kalashnikov or AK47, the most famous and the deadliest gun in the world. So practical and lethal has it proved in modern conflict that it has featured on the coats of arms of Zimbabwe, Burkina Faso and East Timor. There are statues to it in the Sinai Peninsula in Egypt and on the dusty plains outside Baghdad in Iraq. It's had a cocktail named after it and is a drinks brand in its own right, sold in bottles moulded in its iconic shape. Some parents have even named their babies 'Kalash', so deep has been its global allure.

Mark extended a white-gloved hand and pulled down one from China.

'That's a type 56,' he said, putting the barrel close to his face. 'Yes, it's from a northern province. This folding stock was new.' It came from State Factory 66, just one of 15 million of its type produced there since the 1950s. He pulled out another; from 1981, he told me, Chinese as well. His finger traced the first two digits of the serial number. There are about ninety variants of this type, he said, and pointed at a vicious black derivative – one with a hard metal folding stock. 'East German.' He needn't have said any more. It looked East German. There was nothing funny about it.

You could see national traits in many of these guns, however subtle. The Finnish version had a certain chic to it – a tubular stock that evoked northern European woods and candles. The Egyptian one came with a small tree stamped on it, made especially by the Maadi company there on old imported Russian machines. The North Korean one looked cheap and sorry for itself – a small communist star on its base. The Red Young Guard, a force made of up fifteen-year-old Korean students, used this model; it was certainly light enough for their malnourished bodies. Then there were AKs from Pakistan, from Russia, from China – sometimes a dozen from one country alone. There was even an old Viet Cong one – the rifle that proved the ultimate battlefield leveller against the might of the American army.

The one that caught my eye, though, was the gold one: a glittering metal-plated AK designed to commemorate the end of the Iran-Iraq war in 1988. Saddam Hussein handed them out as gifts – a sort of oil-bling chic in limited edition.

'I've got to hold that one,' I said. Something in me was feeling the pull of history, the uniqueness of this whole situation. I wanted to get my picture taken with it, wearing too-large shades and an open-necked shirt.

'Very Arab,' Mark said, and eased it from me back onto the shelf.

He took me to another rack. Here was a Lebanese M16 semi-automatic – this one made by Colt USA. Beside it was an M16 seized from the IRA – complete with a filed-down serial number. Next to it was a line of futuristic and squat black Belgian FN F2000s – a weapon so beloved of Colonel Gaddafi's murderous forces. There was ingenuity to all of these; as you moved along the line many had small modifications that improved on the design of its neighbour.

'If I find a new way of protecting myself, you will find a new way of preventing that,' Mark said.

Then, with a certain reverence, he pulled out an 1805 model of the Baker Rifle – a rifle used on the fields of battle at Waterloo in 1815. It weighed about the same, he said, as the British army's SA80 rifle today: ten pounds. I held it and imagined a scared seventeen-year-old in rank and file clutching its wooden stock with child's hands, fearing all that lay ahead.

We moved away from the military weapons and on to a rack of sporting guns, notable for their provenance and their price. Mark pulled out a hunting rifle – a .375 H & H Magnum – carried by a companion of President Roosevelt on an African hunting trip in 1909. A similar one to this fetched almost $32,000 at auction. Next to it was an M30 Luftwaffe Sauer & Sohn Drilling, the world's most expensive survival firearm. A three-barrelled shotgun complete with a Nazi swastika, it was designed to help Germany's Luftwaffe pilots avoid capture. You could see Hermann Goering's obsession with beauty and craftsmanship in this elegant and totally impractical weapon. And just as I thought that you couldn't get more expensive than that, Mark showed me the most pricey gun in his collection: a bespoke Arab commission of a Smith & Wesson Model 60, made in powder blue and coated in 984 diamonds. It costs over $185,000.

But this opulence was not the thing to catch my eye. Rather, there, nestled in a rack among some ageing rifles, was the prototype for the late nineteenth-century Magazine Lee Enfield Rifle. In army terminology it was the MLE, or 'Emily', the rifle clutched by thousands and thousands of British soldiers as they marched to their deaths in the First World War, and one of the first of millions made. It was also the rifle that had introduced me to the world of the gun – the one I learned to shoot with in the Army Cadets.

Before I could get too distracted, Mark moved us on to a small and nondescript chest of drawers – a cabinet of curiosities. Drawer after drawer of discreet guns used by the secret service were opened. There was the famous James Bond Walter PPK, 9mm;[7] a Parker Pen gun; a 'sleeve' gun designed to be tucked up a jacket; guns disguised as lighters, rings, pagers, belt buckles and penknives. All of them innocuous and all capable of killing.

'Squirrelly,' Mark described them. They certainly captured the imagination – secret agents and honey-traps and the whiff of soupy rendezvous in the fog of East Berlin. We carried on, each rifle catching Mark's eye taken out, examined and admired. So time passed in this space without sun or guiding light. It felt as if we were in a huge mausoleum – a tomb of arms. A feeling of claustrophobia started to form, and the buzzing overhead lights began to hurt my eyes. Then, suddenly, it was time to say goodbye.

I had one final question. I asked Mark about his hand. 'Did you lose it in a shooting accident?'

'No,' he replied. 'I was a Thalidomide baby.' In the late 1950s the prescribing of a pill to combat morning sickness caused hundreds of babies to be born with defects. His false hand had nothing to do with a gun wound. He smiled and bade me farewell and, after another body search to make sure no secret-service pen guns had ended up in my pocket, I left.

Night was falling, and I walked away from this secret vault with its murderous contents, out into a drizzling, darkening northern city. A Thalidomide baby, I thought, turning up my collar. So much for assumptions.

I woke to the sound of an argument. The whores had been up all night, and the dawn was just hitting the sidewalks of Geneva; they were still short of a good night's takings. Such things test the patience of anyone.

I pushed open the window and looked down. The pink neon strips of Le Player's sex lounge shimmied in the lessening dark; the transsexuals, who had pushed their long legs out at the passing men, had left the kerb outside World's Elite hours before, but the women from the Congo were still there. They knew what work really was and they whistled and plucked at the sleeves of men seeking comfort in the early morning. They were sweet-perfumed and hard-faced.

After Leeds I had come here, to Switzerland, to get my facts straight about how many guns there were in the world. The night before I had read that in 2007 it was estimated that there was about one gun for every seven people in the world; that police forces had about 26 million firearms; armies 200 million;[8] and that civilians owned the rest: 650 million.[9] These figures came from the Small Arms Survey – a Swiss-based organisation that lay about a mile away from where I was staying and which was my next port of call.

I had a meeting that morning with their chief, Eric Berman. With almost a billion guns out there, his job was to give some semblance of statistical order to them. He was, in a sense, a worldwide oracle on gun facts and figures. Definitely a man to meet. So I dressed and left the hotel, passed the cat-calling women and headed out to Geneva's waking streets.

The Small Arms Survey was on Avenue Blanc, and the area could not have been more Swiss. The Survey's office was tucked away in the same building as the Myanmar, Cape Verde and Tanzanian missions. Next door to them was a chocolate shop with an oversized cacao bunny in the window. Beside that stood a business school, a medical centre and the Swiss Audit & Fiduciary Services Company. It all had a sterility and orderliness to it that was a world away from the gore and blood that gun violence brings.

I rang the bell. Eric was called for. As I waited, I browsed the magazines on the waiting-room table: *Defence News International*, *Security Community*, *Asian Military Review* – bitter-edged titles. The same could be said of the photography on the walls. One showed bullet holes in a window overlooking a grimy industrial sprawl in La Vela Gialla, an Italian neighbourhood run by the Camorra mafia family. There was a photo of a drug dealer's hand in Brooklyn, clutching a Colt Python .357 Magnum along with fifty bucks' worth of five-dollar crack cocaine wraps. Then an image, black and white like the others, of Liberian youths clutching Kalashnikov-style assault rifles, wearing bandanas. They stared fiercely at the unflinching lens. Guns and their many faces around the world.

Eric appeared, shook my hand and led me to his office. We sat down, and I began to explain that I was writing this book about the world of guns and that . . .

'You don't have to butter me up,' he said, and I was surprised. I thought this was going to be a nice conversation; he looked nice – slim, middle-aged, neat. He reminded me of one of those cautious editors you meet on British papers: clever without eccentricity, focused without shifting into obsessional.

'The Small Arms Survey has many views on guns,' he carried on, answering a question I hadn't asked. 'We don't have a single view. My personal reason for doing this is very different from that of my colleagues . . .' and he began to explain how the Survey is neither pro-armament or anti-gun. Then he stopped, looked at me and said, 'Ask me a specific question.'

So I did. 'How many guns are there in the world?'

But you can't just give a number, he said. Eventually, after telling me how the Survey reviews 193 United Nation member states, and with all the caveats that go with not having access to decent data, he handed me three reports: '875 million was the global estimate.' He then said that number could be higher – this figure was seven years old. He spoke of how secrecy surrounds military and law-enforcement figures, how the Survey has to estimate the number of guns some militaries have by looking at the numbers of soldiers at the height of a nation's power, because when armies downsize their guns are often just put into storage. Such are the challenges of getting a bigger picture. But he did say one thing was certain: more weapons are produced globally each year than are destroyed.

I asked him if counting the number of guns owned by various militaries could help fuel an arms race between countries.

'That's a facile argument,' he said, a spark of irritation deep in his eyes. I was intrigued by how defensive he was being. I told him so.

'It's hard to give you concrete answers,' he said, crossing his arms.

'There's no intrinsic relationship between the quantities of firearms in a given place and the levels of violence,' he said. 'One can really skew one's argument in favour or against gun control. You can pick and choose. You have to be very careful on this topic, as it is so easily manipulated and used. Some people just don't appreciate the complexity of it.'

He saw me glaring back at him over my notebook, and he breathed out. You just have to be cautious, he told me. 'Journalists can take a snippet of something you've said and use it to move an agenda forward, and I don't want to get caught up in that.'

As he spoke, I realised this New Yorker, who had a map on the wall of a hitchhiking trail that he'd trodden years before through the Congo and who had lived in Israel and Kenya, Mexico and Cambodia, was not that dissimilar to me. Guns had propelled him around the world, and his view on them was as shifting as the sandy ground of facts he walked upon.

I had hoped to meet a guide – someone who could have showed me an intellectual and factual path in my journey into the world of the gun. But I'd met someone who refused to be rooted in one opinion, choosing instead ever-changing interpretations offered by ever-changing hard numbers. He told me the world of guns had changed him, that he now looks at data differently and he has to be more cautious in the words he uses to describe his Survey's conclusions. Guns are inherently political, it was clear, and he strived for a consciously impartial voice.

He gradually relaxed and showed me his office. It was filled with softer things: humanity in baubles that had little to do with guns. A paperweight from the Central African Republic, a grave marker from Gabon, a stamp from the Republic of Guinea showing, surreally, Carrot Man from *Lost in Space*. An unopened bottle of Kazakh vodka rested on the shelf.

'Make sure you write down that it was unopened.' And I did, because in the world of guns you have to be careful with the facts, clearly.

After all, it's a matter of life and death.

II. Pain

2. THE DEAD

The gun's mountain of dead in hard numbers – Honduras – the most dangerous place on earth – the tragedy of three murdered women in a jaundiced street – a visit to the fire-marked morgue of San Pedro Sula – witnessing a journalist's trade and a night-time shooting – the secrets of the embalmers' art

Global numbers are hard to come by, but estimates from international studies suggest that between 526,000[1] and 600,000[2] violent deaths happen annually. UN data on homicides show that in areas with high levels of murders, the vast majority of these are with guns – often over 80 per cent of them.[3] An assault with a firearm is about twelve times more likely to kill you than being attacked in other intimate ways, like with a knife,[4] so taking into account that as many as 90 per cent of deaths in conflicts are from being shot,[5] an estimated level of 300,000 homicides with guns every year seems reasonable.

Then there are the suicides. The World Health Organization has estimated that 800,000 people kill themselves every year. As one of the leading ways to end it is with a firearm, a figure of 200,000 suicides by firearm a year also seems a reasonable estimate to make.[6]

This all adds up to about half a million people dying every year from gunfire.

The type of deaths from guns, clearly, differs from country to country. If you live in the US or Canada, suicides account for the majority of gun deaths. In countries such as Brazil, Mexico, Colombia or Albania, the majority of gun deaths are homicides. Eastern Europe and southern Africa have lots of murders, but not many by firearm. Southern Europe and northern Africa don't have many murders, but when they do, they are much more likely to be with guns.

These figures, though, conceal one problem. As Eric Berman told me, there is a fundamental difficulty in getting any figures worth a damn. Many countries don't have proper ways to establish who has died violently, let alone how. Even in relatively developed South Africa, where gun deaths overshadow all other 'external' causes of death, only a third of death records are available for analysis.[7] The World Health Organization's mortality database provides figures for just seven sub-Saharan African countries.

From the data that are available, though, we know that, if you look at the rankings of how people are murdered, Puerto Rico tops the table with 95 per cent of homicides there occurring with a firearm.[8] We also know that Brazil has the most gun homicides in the world outside a war zone, in terms of sheer numbers.[9] And, perhaps of surprise to some, the worst place in the world for gun violence per capita is not the US, but the Central American country of Honduras. And there's one city there that stands out as the world's epicentre of gun violence: San Pedro Sula – the most violent city on earth not at war.

This fact was new to me. I had been to Latin America before – the story of gun violence in Brazil was just one of a number of things I had reported on in the previous fifteen years. This time, though, I felt I had to shift my focus from the United States and instead travel to that heart of darkness of San Pedro Sula, to record what happened to the dead in this city of corroded, wet streets and ivy-curled trees, and to see how people coped under the constant presence of gunfire.

The body was out in the cane sugar field, in the shadows. We stumbled through the night and the plantation mud, the shifting light coming from the mobile phones the police officers were using to guide their way. There was only a weak moon in the Central American sky, and there was no budget for flashlights, so the officials had backed up the mortuary truck and let its headlights cast a low glow across the stubble-rich field. Their phones would have to do the rest.

The call had come over the radio as if it was an urgent murder scene, but the body had decomposed long ago. The sugar cane had since grown and pushed up and out, through the man's jeans. It had pierced his mottled flesh and was now sprouting through his body as if the bones themselves had grown. They looked like lilies in the half-light; you couldn't tell the difference between the bones and the cane.

'See, his hands have been tied,' said one of the forensic examiners in Spanish. He was dressed in a clinical over-suit, but as he was using a garbage bin to put the bones in, it was clear any concern for evidence contamination had long been lost somewhere in the dark corners of countless other crime scenes.

'Is that a rib bone?' The mobile phones were held close to the ground, casting their blue light over the broken earth.

'No. That's a twig,' a voice in the pitch-black said.

'I've found his skull,' said another. An animal must have dragged it away, I thought.

'Looks like they cut it off,' the first voice said. I was wrong. You could make out in the shifting light the ragged hole where a bullet had struck and you hoped they had shot him before they had cut him. Either way, the bound hands and lonely death in a field made it clear this was a gang murder.

This was what I had come to witness, and it had not taken long. I had only been here for a short while, and this was the eighth body I had seen. Honduras, without a doubt, was a very violent place. In 2012, twenty people were murdered every day on average in this country of 8 million – a murder rate of 90.4 per 100,000 residents.[10] In the US it is about 4.7.[11] The city of San Pedro Sula, on whose darkened outskirts I was now, was even worse. The murder rate here was 173 per 100,000.[12] There were, in 2013, just under six homicides a day in this municipal region alone.

The violence was partly down to San Pedro Sula being where it was. Stuck between the drug lords of Colombia and Bolivia to the south and the buyers from the US to the north, it had become a habitat of casual murder and cold pain. Some 80 per cent of the cocaine that reaches US soil was thought to be trafficked via here. And as drugs flowed up, guns came down – from north to south, down from the largest gun-producing country in the world.[13]

These realities, combined with poverty, corruption and impunity, had turned San Pedro Sula into a city where gangs fought gangs and cartels fought cartels over the immense profits that drugs could bring. The feared Mexican syndicates of the Zetas and Sinaloas had even been lured here, aligning themselves with local gangs such as the MS-13 gang or Calle 18. And death had come in their wake.

A few days before, as my plane banked over San Pedro, the lush hills of El Merendon National Park framing the city to the east, and the sprawl of the district of Choloma drifting far up to the north, I looked at my watch. A scattered cemetery speckled the earth in the rushing green below. It was 3.30 p.m. We dipped down to the surging runway. I write this because a skinny policeman was also to note that time – half past three – with a worn ballpoint pen in a crumbling police hill station close to the cemetery I'd just seen. The time was inscribed next to the names of three women who had been gunned down at that precise moment.

The first was Lesley Lopez-Pena. She was twenty-two, single, unemployed. When she died, the policeman noted, she was wearing blue jeans and grey sandals. On the small of her back she had a tattoo of the sun. The second victim was Miriam Portillo. She died with two bullets in her back and one in her chest. The third was Karen Contreros. The report noted that her underwear was pink and that she had five gun wounds in her chest, one in her stomach, one in her shoulder and one in her forehead.

These three women had been travelling back home from a visit out of town. One of them had a boyfriend, a gang member, in prison, and they had been to see him. They had probably given the young man some weed or pills to help pass the dragging hours and then returned. They were caught as they got down from a converted school bus and fell as one from the assassins' bullets. Dying, one dropped a child's Spiderman bike she had bought in the market an hour before.

The policeman did not write down a motive. Murders such as these were just another thread in the endless sorrow of the drug wars.

On the way to the spot where the women had been gunned down, my driver, Frank, had pulled to the edge of the road and put black tape over the telephone number on the side of his taxi. With a deliberate show, he folded a piece of white paper and fixed this over his number plates. He knew the gangs would take these details and he did not want them to visit his home and see that he had a wife and child.

Getting back in, he insisted I lower my window. 'If they can't see in, then they will think we are the other gang,' he said. 'Then they'll open fire.' He was taking no chances.

By the time we reached the crime scene, the light was fast departing, and the coroner's wagon had taken the bodies away. The blood still stained that sandy road, and there was a small piece of intestine, blown out of one of the girl's backs, lying obscenely in the middle of the track. I pushed it with my foot and watched it tremble in the electric light. Perhaps the coroner was too busy to clean up. After all, in the last three years there had been over 6,000 homicide autopsies carried out here in San Pedro, compared to just sixty-two natural death autopsies.[14]

I walked over to a huddle of people sitting back from the road. The mild drama of a Brazilian soap opera was playing out on a square television hanging outside a Portakabin. A fire blazed in an oil drum; the shifting of car headlights illuminated the area and cast dancing shadows. A man in a white England football shirt turned to me.

'Three women?' he said. 'Yes – I heard fifteen gunshots and saw them fall. They lay there for about fifteen minutes before the police arrived, but by then they had been dead for fifteen minutes.'

His Spanish was fast, and because he repeated the word fifteen I was confused.

'The journalists were here before the forensics arrived,' he said, as if that made it clearer, and a fat woman beside him started to scream. I had no idea why.

'The gangs do this as a sort of theatre,' the man in the football shirt was saying. 'They pick where they want the bodies to lie, they leave the gun-shells. They don't care. We have piles of dead bodies here, and the police say they investigate them, but no one gets caught. No one goes to jail.'

The bullets were 9mm. 'Claro'. *Of course*. It's the gun of choice for the feared Calle 18 gang, who run these streets. And with that, he had nothing more to say and walked back into the shadows by his hut. When I approached others they too edged into the dark. The gangs were always here watching. This was just how it was. The killings had brought powerlessness, despair and, ultimately, silence.

Beside us, up a slope, stood a raised breezeblock hut. The lights spilling from the windows captured those inside in silhouette, and then, suddenly, their voices began to lift. They were evangelical Christians. In all of this, perhaps, God was the only one worth speaking to. Below, a line of tied, tired horses snorted in the night, startled at the noise. The cries of those few believers drifted upwards to the speckled sky. And out there, out in the darkness and in an even greater silence, lay three more bodies in a San Pedro municipal refrigeration unit.

Outside the morgue a man in short sleeves and a pair of stained trousers sat and waited and sucked on a bag of fizzy drink through a bent straw. At this time, the sun was already hard on your face, and it would be hours before the heat lessened. The passing cars kicked up small whirls of dust. No one spoke.

Beside him a coffin was propped open with a stick. It lay empty, but he remained hopeful. A quick burial cost about 2,500 lempiras

– $120 – and he looked at the hunched relatives leaving the morgue, with their sallow faces and hurting eyes, and sucked on his straw.

He was from Funeraria San Jose, and was just one of the many morticians who came daily to this, the busiest morgue in the world. It would not be long before he got a customer. His name was Marco Antonio Ramos. At fifty-three, he hadn't thought he would be doing this, but work is work, and this was good work. He had sold six coffins last month alone.

I asked him why he did it.

'Money. I found a way through life with these coffins,' he said, his voice light.

'Do you prepare the bodies for burial?'

'So the relatives can open up the lids and say goodbye to their loved ones – those whose faces are still there.' There are at least ten funeral homes here in San Pedro, and yet business is still good. Just as the lure of death had brought Marco to these gates, so it had brought me – I had come to see how the municipal morgue could cope with so many gun murders.

There was shouting for people from the gate.

'Is there anyone from Baracoa here?' the call went out.

A hunched, fat woman went in, her back contorted, the knowledge of what lay on the other side heavy upon her. Here they got as many as thirty bodies a day; most had died violently. I turned and walked towards the visitor's entrance, the only person to go through those gates that morning with neither tearful nor lifeless eyes.

Inside, Dr Hector Hernandez greeted me. He was the director of this morgue, a tidy man with grey hair and a patient calm, exact and professional. He led me into a large and empty lecture theatre. The walls were peeling, and the place felt like no one had taught here for years. He pointed towards a Formica table and pulled over a decaying chair. Hector's face seemed melted with tiredness. He has a team of 146, he began. Among them are sixty-eight medics, two dental analysts, four toxicologists, two microbiologists and one psychiatrist.

A psychiatrist? I stopped him.

'The morgue is not just for the dead,' he explained. What the gangs do to their victims is sometimes so vicious that their markings on the bodies leave much deeper markings on the minds of those who are left behind. After all, the killers have a method. They almost always end it with a shot to the head – they prefer a 9mm to do this – but they torture their victims first. 'Violence here is intimate, but the gun sends them to the other side,' he said.

Hector sighed when I asked him if this daily arrival of bodies had affected his morale. He was resigned to it.

'In ten years, between 2003 and 2013, we had over 10,000 autopsies; 9,400 of them did not result in an investigation. For me, this is the hardest: this impunity. Nothing has been investigated.'

Right now he had 68 bodies in storage; 48 of them being matched for DNA, the other 20 were unknown. Most had died prematurely and violently.

'After thirty days if no one claims a body, we bury them anonymously,' he said. Last year, 120 people were interred in this way, the majority of them men between eighteen and thirty. Then I asked, in the sixteen years he had worked there, what had stayed with him, what memory of all of this violence had struck him the most.

He sucked in a breath. The murder of an entire family is hard, he said, his voice measured and exact. Like the time he saw a dead mother still holding her three children tight in her arms. The gangs had kicked down the bathroom door and killed them as one. Then there are the others. In this city these are the bodies that come packaged – trussed up in grey sacks. They die painfully, he told me, their legs tied up against their backs, their faces bruised, their teeth missing. They once found twenty-six bodies in sacks like this in a field: a grim harvest.

Suddenly, as if this was too painful a memory to dwell on, he rose, straightened his tie and beckoned me to follow. We walked through swinging double doors and out into the dissecting room. It was a sudden shift from talking about death to seeing it.

The tiles on the floor were loose and covered in water. The neon lights gave off a sickly glow and buzzed; the walls were smeared and wet. And there, on the left, lay a body placed on its side. It – he – was naked, and his legs were crooked and twisted. He had been shot in the jaw, and flies flickered above him.

The director leaned towards me in the molasses air and said there was no real danger of infection. 'The dead are healthy. They didn't die from diseases.' Later, I walked outside and saw bags of seeping waste left against a wall, frenzied flies thick above the trailing lines of blackened ooze, and was not so sure.

We left and I followed Hector upstairs. A fire had ripped through half of the morgue on a summer's night a year before and now the upper floor lay derelict: tortured iron railings and marked walls. Such is the state of Honduras's morgues. As if death had seeped into the very structure of this place and left it rotten and mould-tainted.

Later, he introduced me to his medical colleagues. They shifted in their blue shirts when I shook their hands – they were embarrassed to be asked questions about what they did. Their work was difficult, Hector explained, and I asked what sort of people were drawn to this type of task. He repeated the words of the funeral worker outside: there is not much other employment around. Death creates its own labour.

I offered the coroner team something to eat, and we sat down together. Around the table were Sanchez, Garcia and Rodriguez, two doctors and a forensic photographer. I had bought fried chicken and, despite the sugar stench of death coming from just beyond the door, they ate their lunch. I did not; I had gone to the toilet to wash my hands and found neither soap nor towels.

I asked about the smell. There was a smirk. 'What smell?' These men had been busy and were hungry. On the day before they had nine bodies brought in: six homicides. Outside lay two more bodies. I looked at the white chicken meat and fried strips of skin in their hands and focused on writing notes.

'Look at this. This one has been shot in the head,' said the forensic photographer, glancing at the laptop before him, his mouth full. I shifted across to his screen: it was one of the women who had been killed the day before. There was the child's Spiderman bike. The doctors looked too but were unmoved. The only thing shocking, they told me, is working with children who'd been tortured. One of them let out a low whistle. 'It's really common.'

Luck, fate. These were the things they talked about – as if that's all you could pin your hopes on. 'Some people are shot twenty times and end up in hospital, still living,' said Sanchez, a heavy-set man with eyes dark rimmed and deep. 'Then there are people who are only shot just the once – a small wound – and they end up here.'

'The beautiful thing about this job,' said Garcia, wiping his fingers with a napkin to clean off the chicken grease, 'is seeing up close what a bullet can really do to you.' And then he picked up another chicken leg.

That night I met Orlin Armando Castro – a local TV journalist with a fixed gaze and an impish laugh. He had a fizzing energy that meant he never stopped moving. Beside him was his cameraman, Osman Castillo, a solid man in ripped jeans and a white shirt. Osman hardly spoke; Orlin was his voice.

On Orlin's belt was a police radio that buzzed from time to time, and in his hand, always, was a Blackberry phone. He constantly scanned both and replied to his messages with a focus that could have been mistaken for something else. He was constantly awaiting that call – to a murder scene, to another death. On hearing of one, he and Osman would jump into their scraped blue Hyundai Tucson, whose passenger door did not open from the outside, and drive fast to where a body was sure to be lying. There they did what they were paid to do: they filmed murder.

I had arranged to meet Orlin because he was a local journalist here and I had been told – out of everyone – he was the first to get to San Pedro's murder scenes. The one reporter the police would call whenever there was a shooting, his life was defined by gun killings. And I wanted to know what that could do to a man – to be a constant witness to the tortured secrets of this city, to have a career marked so powerfully by the gun's ultimate legacy.

It was late when we met outside the chipped and long-shut-down hairdresser on a darkened corner of a crossroads. We shook hands, and then, casually, Orlin pulled open his car door and showed me his guns: a 12mm shotgun and a 9mm Beretta pistol.

'Have you used them?' I asked him in the half-light.

'Yes,' Orlin said. I wasn't used to journalists packing heat, less so firing them. One time, he said, he drove into a gunfight by accident. He had to put down his microphone and pull out his pistol and start shooting, because the gangs, in the confusion, had begun to shoot at him. Even so, he refuses to wear a bulletproof vest because the gangs might think he's a cop and then they'd be sure to kill him.

He had worked for the past eleven years for a national Honduran news channel, Canal 6, and had seen things on these eternal, yellow-lit night streets that you should not see. A six-month-old killed in the middle of a gunfight; whole families executed in their homes. He looked at me, his head tilted slightly, and flipped around the screen of his white Blackberry phone. On it was the decapitated body of a woman, her vagina on display. His thumb flicked, and another image appeared. Three day-old dead men lay in cornfields, the heat causing their eyes to pop out of their heads. He laughed, his eyes twinkling, and he showed me another woman, semi-naked in death. His phone was filled with corpses. Young men from the 18 gang slumped in awkward positions, as if asleep. Before and after shots of the living and the dead, from smiling to something else.

When he does not work, he gets bored, he said. There's so much drama in what he does. The closer he gets to death, the more alive he feels. This, he told me, was real journalism. I began to fear this little man's love for the tenebrous corners of this city.

There's much that he cannot report – if he did he'd be killed. Some murder scenes he just has to stay away from: he knows things would get too complicated with the gangs if he reported on certain killings. He feels he's walking on an edge. 'On the one side there is deep, dark water, on the other side there is fire. Here you don't know who is who. In a war you take sides. You know who an army is – they are in green. But here . . . you have no idea,' he said.

A call came in. There had been a shooting in the Barrio Rivera Hernandez, and Orlin's face changed. We jumped into his car and we were off, pushing through the down-lit streets to the murder scene. In this light the street took on the colour of jaundice, the plaster on the low-slung houses hanging like pockmarked skin, the grill-lined windows the shade of mustard gas.

The body lay still under the ash-blond glare. The policemen were placing small fluorescent triangle markers out under the shadowed light, tracing where the spent rounds had fallen. The body lay awkwardly, his legs twisted, the shoulders tucked underneath. The dead man was wearing an orange polo shirt, which looked almost white now, and you could glimpse tartan boxer shorts poking above his stained blue jeans. When the cameraman turned on his light, you could see the blood still seeping gently from the man's back.

The police took out a tape measure and began to measure the ballistic range, but you felt they were doing this because the television crew was nearby. The police spoke to no one, and the street's occupants stood back in the shadows. All the neighbours had come out to look and to talk in quiet voices. A fat baby sat on the sidewalk, gurgling; a girl, about three years old, in a pink frilly dress with small pierced ears, asked her mother for a hug; to her side a man laughed and swung his son between his legs. And in front of these children, the police flipped the body, and the man's destroyed face stared up into the deep black sky.

Orlin, his face caught in the camera's brightness, stood before the body and delivered his lines, repeated a thousand times before. And the image on the video screen showed him, the whiteness of the light hard contrasting with the sulphur-tinted streets, like a broken angel. Luminescent. Then the camera's light went out, and Orlin turned and took one more photo with his phone, and another crumpled face of death was captured.

When they finally put the dead man into a long, rustling black bag, the crowd grew bored and drifted away: the show was over. And the police tipped the body into the back of the forensic truck and then they too left; and all that remained were patches of sticky, coagulating blood, thick on the ground.

Orlin walked back to his vehicle. I caught a glimpse of his face lit in the reflection of his phone. He was looking to see if any more murders had been called in that night. And so it goes, I thought. The endless hunger for death in these streets never sated – one that totally consumed this slight, sad-faced man. I climbed back into the car and we drove away.

The low barbed-wire-rimmed walls of the district flickered beyond the window. And the silent homes of the people of San Pedro, with their contained patches of blue electricity, began to thin out, until all that was left were the spotlights of the car and the silence, and the yellow streets in the rear window diminished into the night.

The coffins attached to the wall are the pricier ones, Daisy Quinteros explained to me the next day, pointing to the far end of the funeral parlour shop.

'The most expensive is 54,000 lempiras,' she said, smiling – just shy of $3,000. She was a good saleswoman and dressed appropriately for this sad room: motherly. Her hair was flecked with lines of white, and her trousers a smart grey that strained slightly around her hips. She wore a tastefully embroidered white shirt. The look clearly worked – she sold about three coffins a week, getting a commission from each. She once earned over a thousand US dollars in just one month, she said.

We were overlooking a street lined with funeral homes. The kerbs were filled with solemn cars, and beside them pine trees cast spots of shadow onto the baked pavement. One of the funeral-home owners had planted white, almost translucent, orchids in pots leading up one stairway; and all around the entrances and pavements were swept clean. Unlike other parts of the city, this area was free of graffiti. This street looked the richest of them all.

I had come here to see one more community impacted by the gun – to look at the art of the undertaker. In San Pedro you did not have to travel far to meet one.

Daisy beckoned me to sit down at the glass table in the centre of the showroom. Unusually around here, she had not lost anyone personally to the violence. That was not to say that it had not affected her; the suddenness, the shock of death coming unexpectedly, these were the things that still disconcerted her.

'You can see it in the eyes of the family members,' she said, and leaned forwards and touched my arm; 90 per cent of her clients had died violently.

'It's not all bad, though. The other day we buried this old man. He was 102. No one lives that long here.' And she smiled a thin smile, because she knew this wasn't what I was here to write about.

I asked her if earning a living from the violence bothered her.

'Well, we've been here twenty-one years. We provide a service – we are a necessity. I don't think our business is taking advantage at all. What would they do without us?' She talked quickly and without pause, her moving hands covered in gold rings. 'Everyone is going to need this service some day.' She pushed a folder towards me. It was filled with images of coffins and garlands, plaques and headstones: a catalogue of death.

'So – how would you like to be buried?' I asked, and through the tinted windows you could see a chain of cars pass slowly outside. Another cortège. Another profit line reached.

'I'd like a mid-range coffin. I've already bought it.' She flicked through the laminated sheets and pointed to the one she had in mind. It was modest, and beside it was a list of measurements. People are getting fatter, she said, now you have coffins in XXXL. But they only come in a set height, so with a 6ft 2in. man like me they would have to do something to reduce my leg size. She didn't elaborate, and I imagined someone shortening me with a hacksaw on a metal gurney.

Daisy seemed the happiest of all the people I had met so far in this city. Perhaps her job was meaningful in a way others were not. She still had contact with the living – even if they were suffused with grief. Other professionals I had met in San Pedro, like Orlin, had jobs that focused on the bodies delivered by the carnage. But Daisy dealt with those with breath still in their lungs. She had to be professional and sympathetic, not least to help families navigate their way through the layered choices presented to them in her laminated folders.

Later, I sat down with Daisy's hidden counterparts: three embalmers who were brothers. They were in their fifties and had the same triangular and light-brown features. One had lived in the US for many years, and the good living had bloated him to twice the size of the others, but they all had the same eyes. Eyes that had seen things get steadily worse over the last five years: 'Once we buried five people from the same family, all dead from guns,' one said. 'We prepare far too many teenagers for the ground,' his brother added, and the three nodded in unison, like priests. 'Many are just fourteen years old,' the third said.

Their skill stretched back to their grandfather, ninety years before. It wasn't like it was now, not back then. But theirs was the oldest outfit in Honduras, and they were still working hard; on average they prepared thirty bodies a week. The preparation took place out in the back, away from the light of the shop front.

They led the way. Past a line of neat walnut-coloured coffins, through heavy swinging doors and out to a room that looked like a cheap operating theatre with a metal trolley at its centre. But here there were no machines to monitor life: just things to mimic it.

To the side was a kitchen tray bearing lines of mascara, rouge, lipstick in neat, ordered rows. In this Catholic country, the casket was often left open at the funeral. People wanted to file past to bid farewell; death was so often sudden and unexpected many things left unsaid had still to be said. So these brothers worked to make sure that the bodies looked peaceful. They erased the look of terror imprinted on lifeless faces. They brought back the illusion of serenity – peaceful resurrection with a make-up bag.

The eldest, Arnold Mena, a softly spoken man in a crisp white shirt and a lined jacket, was so good at what he did that it wasn't an issue if you'd been shot in the face. 'One shot, two shots, three shots – as long as the bullets don't destroy the face – you can just stitch up the entry hole and cover it with foundation.'

'Here they use smaller-calibre guns, and that doesn't break the face so much,' Arnold said. 'But if the skull is totally destroyed . . . we have to use a small football to keep the shape.'

He explained how they use small prosthetic eyeballs too, but then they have to keep the eyelids closed and fix small pins to keep it all in place.

'The real challenge,' he told me, 'was when we do not have a photo and do not know what the victim looked like. Then you have to be a little creative.'

They did other things here, too. In that stark room, beside a metal table with an ugly drainage hole for the dripping fluids, stood rows of formaldehyde from 'The Embalmer's Supply Company'. Twenty-four bottles cost $180 here, and that was enough for twelve bodies. 'It will keep a body for a week, even without refrigeration,' they said, even in this Central American heat.

Beside the bottles were small plastic bags. They put the intestines inside these. The bags were then sent elsewhere to be burned, and they packed your body with 'pulverised hardening compound' instead.

After a while I shook their hands, and they told me to stay, to come back soon, but I wanted to leave. I did not want to know more about plastic bags filled with intestines or skulls filled with balloons. And the smell had long ago seeped into my clothes.

I had seen enough of death's ugly business – I knew all too well what the gun could do. I just wanted to head back to the land of the living. Or, at the least I wanted to see a glimmer of hope in all of this sunless despair; so I left and sought instead to meet those who had managed to survive the gun's barbed impact.

3. THE WOUNDED

South Africa – a bedside visit – the gun's hidden impact revealed – a chat with a trauma surgeon – a blood-tinged night in a Johannesburg emergency ward – understanding how science feeds off the gun's misery – a trip to the BBC to meet a paralysed correspondent

The boy – for he was hardly a man – lay there and watched me. His chest rose and fell, and my eyes drifted from his handsome face to his stomach, where he had been shot.

It was a hellish place to be hit. The bullet had ripped through his intestines, leaving a gaping and ragged hole. Five weeks had passed since the rushed horror and blood-soaked panic of that night, and the wound still refused to heal. The shit from his bowels was re-infecting the coarse edges of torn flesh, and you knew this because of the stench. The doctor spoke to him quietly in Afrikaans. He was eighteen, and there was a chance he would have to carry a colostomy bag for the rest of his life.

Three other South Africans lay in that room. Each shot. Another six lay in the room next door. These men, too, had been shot. And in the room further along another six lay. By this point you had stopped asking the doctor what had happened to them, whether they had been shot, because this was Cape Town, and this hospital was the main medical centre for one of the largest townships in South Africa. And as such it was home to one of the busiest trauma units in the world for gunshot wounds. Which was exactly why I had come here.

The room was empty except for their beds. No flowers, no cards. One man turned in his delirium and moaned; his back was sweated out, and his head was swollen. His breathing came in low gasps.

There were fifty-one beds in this trauma ward, and sometimes it got so bad the overflow spilled into the waxen, squeaking corridors. There were only four nurses on staff, and that was never enough. Last month, three men were brought in. All were in the back seat of a taxi when someone had fired a single high-velocity shot through the car. They were all hit by that same bullet – six legs to be treated, the round clean through. And that story alone filled up three beds just there.

The ones who lay here, sullen now the pain had passed, were young men. More would come. Tonight was a Friday night, and the weekend brought in the bodies. And I looked at the plastic bag that had been taped over the young man's stomach and wrote something in my notebook that I later was unable to read.

What this scene reminded me, like the dramas played out in thousands of wards in slum-towns and war-zones the world over, was that the majority of people go on to survive being shot.

It works out as a hidden epidemic of pain and violence. In the US, up to 91,000 people were admitted to hospital with non-fatal gunshot injuries in 2011 compared to 8,583 who were killed in shootings.[1] In the UK, it is estimated that 777 people were shot and survived in 2012,[2] compared with about 150 killed by gunshot the year before.[3]

Such harm is hard to imagine, but consider this: about 35,000 American children and teens are said to have suffered non-fatal gun injuries in 2008 and 2009 – six times more than those shot and killed. This is the equivalent to 700 school classrooms of twenty-five students each: a number greater than that of US military personnel wounded in action in Iraq and double the number wounded in Afghanistan.[4]

Admittedly, the exact numbers might be debatable, but what is not challenged is that each injured child experienced the horror of a bullet crashing and crushing through them. Their tissues and bones and muscles were shredded in the bullet's path. Their insides were horrifically displaced, as if kicked by a mule. Their bone fragments spun off and lacerated and pierced their young bodies.

Ultimately, these children's chances of survival were dependent on a host of factors. The speed of their getting treatment was crucial – the so-called 'golden hour'. In one study in the US, it was found the likelihood of you dying from a gunshot wound was about 25 per cent higher if you were shot 5 miles or more from a trauma centre, and you could not get there within the hour.[5]

Also important was the wealth of the country in which they were shot. In the US for every person shot and killed, as many as nine survive. In developing countries the ratio is far smaller; more people who are shot will die – about one in three. The World Health Organization estimates that between 50 and 80 per cent of traumatic deaths in developing nations happen before people get to hospital, in part because, in many areas of the world, ambulances are almost non-existent.[6]

Your chances of survival also come down to factors far beyond your control. The bullet's weight, the speed at which it hits you, even the pull of the moon has an effect. It is all about the transference of kinetic energy in a chaotic way; variables that determine the final resting place of the bullet or how badly you are hurt are all unfathomable.

Other things matter. If you're wearing clothing at the time, there's a greater risk of damage and infection to your body.[7] If you are pregnant there are sometimes significant complications.[8] And, in the US at least, whether you have health insurance plays a factor. One study said uninsured trauma patients were more likely to die after being shot than those insured.[9]

It is not just the immediate trauma of the wound that causes harm. Bullet fragments left in the body can also result in higher blood lead levels.[10] Or you can go on to develop related health concerns – as in the case of US President William McKinley, who earned the title of being the first reported case of traumatic gunshot pancreatitis.

All in all, getting shot is a terrible lottery. The odds might be in your favour, but it's one bet never worth taking. Thank God, then, for doctors.

In a closed room behind a steel-barred door the medic and I sat and talked. The room was lit with an ugly sterility, the overhead lights gave off a low buzz, and all around was chrome and glass, instruments wrapped in stark, sterile packages. A bleach-white smell clung to this place. For two years, Dr Taylor, a petite and vivacious young woman, had been the head of the trauma unit here in this rising, brick-built oasis of South African care: Tygerberg Hospital.

Tygerberg. It sounded like the tiredness and despair that had long ago infected the slums surrounding this place. Each month up to 2,000 patients passed into Dr Taylor's world, fresh from the poverty of the Cape Town flats. And what she saw, endlessly, was the trauma wounds of penetration – gunshots and gunshots and gunshots.

'In the past we got stab wounds, but now it's gun wounds. It's all to do with drug crimes and gangsters.'

She was thirty-four. One of those bright young doctors whose sparkle gives you faith in this world, one who had always wanted this sort of work. She was from South Africa's Free State and had that matter-of-fact way about her that defines people. But these gunshots, this was new to her. Back home, back in Bloemfontein, a place of long pastoral lands and the languid time of rural life, it was all stab wounds and car accidents. Not like here in the Cape Flats.

We were in the controlled section of the hospital. Only the staff and the dying and those clinging to life went back here: an unseen world the gun helped create. A world of sterile swabbed pain. We had walked through clanging doors and down long lines of scuffed corridors that glowed in the off-white light and turned into a windowless room. There we sat at a metal desk surrounded by blood pressure gauges and ventilators, IV lines and machines whose purpose you didn't want to know. Drugs lay in quick-grasp handfuls in cabinets that hung upon the scrubbed walls. Adrenaline, Etomidate, Furosemide, Atropine – alien and painful-sounding words. There were ugly things that caused pain. Scissors. Scalpels. Large-bore catheter needles, sixteen gauge. They spoke of one thing: that the pain caused by guns does not end with the pulling of the trigger. That's just the start.

Her patient population was predominantly, almost exclusively, young black and coloured men. And the gunshot wounds were predominantly low-velocity and multiple. No AK47 rounds here; rather, small hand-guns and bang, bang, bang. People getting shot four, five times even.

'One guy was shot thirty times,' she said. 'Mostly flesh wounds, but he survived.' She tapped the desk. She was frustrated with the lack of resources. She wanted to help so badly, but things were never just about desire here. 'In the US you have full body scans. Full diagnostics, all on hand for you there. All in fifteen minutes. But here – we see so much violence and we've only just got an ultrasound.'

Computerised tomography scans and X-rays here can take twenty-four hours to get back, and this made for hard decisions. The other day a patient came in – shot in the abdomen. They put forty units of blood and blood products back into him, but by then he had suffered renal failure, so he went on a ventilator and then into an intensive care unit for four weeks. This meant others were refused intensive care, there just weren't enough beds – and by others she meant children with acute appendicitis or cancer. One life saved here, even if it's the life of a killer, means another life lost.

This is the stark reality of trauma surgery in a land of scarce resources. The ratio of public doctors to patients can be as low as 3 for every 100,000 in the South African health system.[11] Such state medics care for about 85 per cent of the nation's trauma cases and these men and women in white clearly don't have enough resources to cope.

The gun has hardened her, she said. No longer does she want to be told about the background of her patients. 'I am not interested in knowing anything more than that they were shot. I don't need to know that one guy, a guy we've spent a long time helping and given lots of resources to, that he then brags about how many women he has raped. That's too hard to hear. I don't want to know, because, you know, lots of them have raped and killed.'

Her voice slipped a little in the white room. I noticed an edge of anger. I suspected that she, her heart so full of care, couldn't comprehend how others did not feel the same desire to change matters, to help. But the thing she found the hardest was that those whose lives were marked by violence – the gangsters, the young thugs – often survived the terrible wounds caused by guns. It was the passer-by – the innocent caught in the crossfire who never expected this – who died with a look of surprise on their face, unprepared for the sudden descent. That bothered her.

She had seen a change in her personality. Now she is more clinical, dispassionate. 'Don't come and cry on me.' This was what this brave doctor said to people. And she looked a little guilty at what this had done to her relationship with her patient boyfriend, an engineer and a man who never had to hold a dying teenager or an infant with a gaping gun wound in his back. Other things slip, too. After days of bloody surgery, everyday chores like tax forms and bill payments and driving licence renewals just fade out. Death captures her attention like a demanding child.

'There are days when nothing happens. Then a whole number of gun-trauma victims come in at the same time. It's 0 per cent to 100 per cent. In those days when nothing is happening, you pace the corridors; you get bored. You find yourself only functioning when something happens – when you are on adrenaline,' she said.

She had been soaked in blood, head to toe, several times in the last year alone. So I asked her about HIV in this land where about 10 per cent of the population are infected, and her answer was as brutally logical as her other answers. She didn't think about it – she took the necessary precautions with double gloves and all the rest. But it's not possible to avoid blood. If there was cause for concern, she would take antiretroviral drugs and to hell with it.

'They aren't good. They make you tired, give you diarrhoea. You vomit. So you ask yourself – what are the chances of getting an infection? You treat all patients with caution but you can't discriminate.'

Blood was nothing to her. But, then again, she couldn't watch horror films. She was scared of the dark.

I ask about what would be the worst type of gunshot, and she was quick with her answer. 'The head. If the head is involved and the bullet has gone through – well, it's a very poor prognosis. If vasculature is involved, if you get shot in the neck, chest, abdomen and it is close to a vessel – all of these have poor outcomes.'

'In fact,' she said, 'we don't get to see a lot of large vessel abdominal injuries because those shot there just die.'

If you are shot in the limbs, she went on, you can get devastating trauma to nerves, or you get complex fractures, and then young men lose their legs. But what is most horrific is a spinal injury: C3 fractures, quadriplegics, tetraplegics. They end up in care homes and lie there, and no one turns them. No one cares for them, until their own foreshortened death.

She descended into talking about sepsis and perianal wounds and genital trauma. But her mind drifted back to those she felt most powerless about. It was the bleeders that stuck in her mind, those shot in the portal vein, the retrohepatic inferior vena cava, the aorta. They die there on the table, and you ask yourself: 'Did I do the best I could?' Holding these nameless men as they slip into unconsciousness and beyond has meant she has begun to take sleeping tablets to help her sleep. Or she turns off her phone and goes for a run and just, well, just tries to live a life of the living.

'In the end,' she said, patient and calm and answering my questions as best she could, 'you really just want people to survive.'

The man's face had the look of wax; his eyes were glazed and unfocused. He had sustained a vicious beating, and it was unlikely he would survive the night. His leg moved in small, grotesque, primal jerks. The man beside him was breathing in short, sharp gasps. That one making urgent noises with a bloody drip coiled up and away from his chest had been stabbed with a screwdriver.

The night after my conversation with Dr Taylor, I had driven back through the streets of Cape Town, through patches of contained light cast upon the empty dark roads, to witness what a weekend night

brought to this hospital. To see people on the edge of surviving and
to see which way they'd fall.

On this midnight watch, the waiting area outside Tygerburg's
trauma unit was filling up. A small boy lay silent, supine in his
mother's arms. His thumb had been ripped off, and the nurse was
telling the mother they would not be able to save it. Later, the mother
was to ask me if I was able to help him, because I was white and
she assumed I was a doctor.

The paint was coming off the ochre walls in thick strips around
the four ugly hooks that hung there. They were for saline drips; the
numbers of wounded here was so great that no space was left inside
the unit.

A sixteen-year-old walked over and sat next to me. He had been
stabbed in the neck over a 100 rand payment – about $10. His
mother sat opposite. It was the first time he had been stabbed, and
he laughed when I asked him what he was going to do about it.

'Payback,' said another man. The boy with the ripped thumb
drifted to sleep.

A consulting doctor came over and talked to the boy and then
turned to me. This young medic had been here ten hours already,
and he'd seen things, he said. Like when a man had come in with
six bullet wounds in his knee, and when they raised his thigh to get
a look, the rest of his leg had just stayed on the table. Or the one
who had had the top of his head cut off with a buzz saw.

He led me to the doctor's area – a quiet room at the back behind
a scuffed door and away from the noise of those in pain. Inside were
other doctors, huddled in close, like fishermen sheltering from a
storm. One was from Switzerland: a handsome man who had travelled
to over eighty countries and whose girlfriend, who once skied profes-
sionally, was also a doctor here. They were an impossibly attractive
couple in this ugly place. His words tumbled out; in trauma units
time is of the essence, and there is no space for languid talk.

If you are a trauma surgeon, you don't want to work in a quiet
hospital, he said. So, you come here to see what guns can do, for
there are few other places like this in the world. Doctors like him
come from Holland, Sweden, the US, the UK. Some have never seen

such penetrating trauma. An eighteen-month-old hit in the crossfire. A mother raped and shot as her two-year-old played beside her. These doctors had learned much. Like how to drain a heart with just a needle, or perform three laparotomies in a row, or hold a dying man so he did not go into the darkness alone.

'Without a doubt,' another said, a big man in a white coat and a solid voice, 'South Africa is a violent nation. It's like a civil war. I've spoken to guys in Iraq and it's like this here on a Saturday night.'

Then an emergency call came in, and they solemnly filed out, back into corridors swathed in dull electric light.

I was left alone, and I thought how the gun had transformed these medics. How it made them stronger surgeons, more confident, more able. The harm that firearms wreak had caused them to develop skills and tools to bring people back from the edges of life. And they, unlike the men and women I had seen in the morgues of Central America, could offer hope in a landscape of despair and death.

A pile of papers lay to one side, and I picked one up – a medical magazine, *Trauma*. Its reports were revealing.

> Initial surgical management of a gunshot wound to the lower face.
> Non-operative management of abdominal gunshot wounds.
> The European Trauma Course: Using experience to refine an educational initiative.

The last title showed just how much the trauma community is tied together by a singular response. Bearing witness to horror, they must learn from it. And this impulse to learn has transformed the course of medical history. For without learning from the history of the screams of men like the ones who lay shot outside this room, the gun truly would have won. Else it would have only taken and not given back a single thing.

War and violence have been the engines of creativity for many things that we take for granted. A material called Cellucotton, for instance, first used in the First World War to patch up gun wounds, was so absorbent that it caught the nurses' eyes, and the sanitary towel was invented. The Great War also saw the creation of, or at least popularised, the tea bag, the wristwatch, the zip and stainless steel. But war, most pointedly, has been a constant driver of medicine.

As guns have evolved through the centuries, so too have medical responses to the injuries sustained from them. And the injuries have been terrible. In the fourteenth century, gunpowder's arrival onto the battlefield made the treatment of trauma wounds far more complex. No longer the splice of a sword or the pierce of an arrow. Rather, embedded bullets, gunpowder burns and gaping holes in flesh changed forever the nature of wounds.

The early modern doctor was ill equipped to deal with such complex trauma. For a time gunpowder's ability to take life so easily was even put down to the belief it was poisonous and that bullets were contaminants. This led to the medieval practice of burning the wound to rid the body of poison.[12] Of course, such treatment probably took more lives than it saved, but it was not until the mid sixteenth century, when the French military surgeon Ambroise Paré, in the thick of battle, ran short of hot oil to cauterise wounds, that anyone challenged this practice and, more importantly, wrote about it. Paré improvised: egg yolk, rose oil and turpentine were used instead, and the benefits were marked. Many more survived under his care.[13] But innovation takes time to find roots, and the technique of pouring boiling oil into wounds continued for another 200 years.

Bloody death after bloody death, though, has a horror that cannot be ignored, and the impulse for doctors to learn and to understand remained. Clearly much of that was by trial and error. So the American War of Independence in 1775 saw the surgeon John Hunter suggesting that, if a gunshot wound was to be sewn up, a piece of onion was best put inside, and then the wound reopened after two days. The presence of pus following this procedure was seen as a good thing: a sign of the wound healing. But during the Crimean War in the 1850s, a connection between mortality rates and sanitation was to become firmly established. There Florence Nightingale 'was to thoroughly scrub the hospital, provide clean bedding, improve ventilation and sewage disposal', with notable impact on patient mortality – it dropped almost immediately from 52 per cent to 20 per cent.[14] This vicious war also saw the widespread use of chloroform to alleviate pain, and plaster of Paris to treat bones shattered by grapeshot.

But, just when it was thought medical discovery was catching up with weapons technology, the Minié ball came along. The round balls used before this tended to remain lodged in the flesh and muscle. The Minié ball, on the other hand, cut straight through, leaving a gaping, haemorrhaging exit wound; the metal rarely remained in the body. If a Minié ball was to strike your bone, it often caused it to shatter, causing damage severe enough to require amputation. It turned mass infantry assaults into mass slaughter.[15] Fatality rates shot up; penetrating gunshot wounds to the abdomen reached a mortality rate of 87 per cent. There were over 50,000 amputations in the American Civil War, and infections followed, the spectre of death hard on their tail.[16] Tetanus had a mortality rate of 89 per cent and pyaemia, a type of septicaemia, killed 97 per cent of those who developed it.[17]

So devastating were these odds that, by the Spanish-American War of 1898, the medical profession recognised the urgent need for antisepsis. After reading findings by Louis Pasteur, Joseph Lister carried out experiments using carbolic acid and found it helped massively reduce the patient's chances of dying if applied following amputations.[18] Antiseptic dressings on the battlefield and saline solutions to hydrate patients were also brought into play – innovations conceived on the bloody, ragged fields of war.

Roentgen's discovery of the X-ray in 1895 further revolutionised trauma medicine. In previous wars, unwashed fingers and metal probes were shoved into screaming men to locate bullets and metal shards. Lost pieces of cloth could be lethally dangerous, suppurating and causing gangrene to topple a man, but the use of X-rays in the field helped pinpoint fabric, bullets and bone fragments. The need for amputation and the subsequent risk of infection were greatly reduced – so much so that what happened to the mortality rates of the US wounded in the Spanish-American War was nothing short of revolutionary: 95 per cent of wounded men recovered.[19] It was a far cry from the carnage that had defined the American Civil or Crimean Wars.

Then came the First World War. Those fixed lines of carnage brought their own rat- and slime-filled horror, but they also meant that those who were not caught dying upon rusting barbed wire had a fighting chance of survival. The rapid evacuation of casualties from the front line massively improved a wounded soldier's chances of living. There was a mortality rate of 10 per cent if those hurt were casevaced within the hour. If you were out in no man's land for eight hours, your chances of death rose to 75 per cent.[20] The Great War also saw the wide-scale use of the tetanus antitoxin, and deaths from lockjaw dropped from 9 per 1,000 wounded to 1.4 per 1,000.[21] But perhaps the most significant medical innovation was the first blood bank, established by Captain Oswald Robertson in 1917.[22]

The Second World War added to this: the development of blood banks continued through the early 1940s, as well as the rapid evacuation of the wounded and the production of penicillin on an industrial scale.[23]

By the time the Korean War began, things had improved beyond recognition. Casualties were being evacuated by helicopter, and plastic bags had been introduced to replace the glass bottles used to transport blood for transfusions. The conflict also saw the development of mobile army surgical hospital (MASH) units, which brought surgeons to the front lines. They were literally life-changing. A wounded soldier who arrived at a MASH unit had a 97 per cent chance of survival.[24]

Medicine and surgical techniques have continued to keep up with modern warfare, and their benefits have been passed on to civilians. In 2013 a trauma centre at St Mary's Hospital in London launched a new medical process based on a protocol developed at the British Army's Camp Bastion in Afghanistan. It was a triage system to treat gunshot patients as quickly as possible, taking casualties straight to surgery to stem the bleeding.[25]

Other recent medical innovations in gunshot trauma include the drug Tranexamic acid. In 2010 a study into this drug, originally used to ease heavy menstrual flow, showed it could save the lives of haemorrhaging patients.[26] The drug was quickly adopted by the British and US armies and is now seen in many American emergency departments. There has even been the development of syringes containing tiny sponges that can seal a gunshot wound in seconds.[27] The reality of gunshot victims being placed in suspended animation, or 'emergency preservation and resuscitation', is also upon us. This involves replacing all of a patient's blood with a cold saline solution, which rapidly cools the body and stops the majority of cellular activity, giving doctors time to treat the wounds methodically, without the tick-tock urgency of a dying patient on the table.[28]

But what all of these medical advances mean is that we cannot view the impact of guns solely in terms of the numbers killed by them. Given so many people are now being dragged back from the edges of death by the medic's steady hand, we have to factor in the numbers wounded by them as well if we are truly to understand the gun's impact.

The BBC foyer was filled with day-trippers. A coachload of excited, heavy-set tourists were down from the North, full of laughter, teasing each other gently. Some were having a go at being newsreaders in an 'Interactive Newsroom' corner. A huge poster of Annie Lennox dressed as an angel looked down. I sat down on a puce kidney-shaped sofa and thought about the person I was to meet: Frank Gardner, the broadcaster's diplomatic correspondent.

Ten years earlier, Frank had been gunned down by six Al Qaeda thugs in Saudi Arabia. He had been shot a number of times – in the shoulder, leg and, at point-blank range, four times in the lower back. His colleague, the Irish cameraman Simon Cumbers, was killed beside him. Frank had lain there, in a spreading pool of blood, for the better part of an hour until he had been delivered, as minute seeped into agonised minute, into the capable care of a surgeon who had worked in the very South African hospital where I had seen those trauma victims. The training had, clearly, been of use, and

Frank survived. Just. The bullets had missed his major organs. But one had clipped his spine and left him partly paralysed in the legs and dependent on a wheelchair. That was why I was there: because of his pain.

We are seduced by the idea of the wounded poet. The warrior hurt beyond hurt, yet a hero who, against the odds, rises through agony, overturns death and emerges, filled with knowledge, into the light. Perhaps I imagined Frank like this. After all, since the shooting, he had been given a medal by the Queen and written two bestselling books. Following fourteen operations, over half a year in hospital and months of rehabilitation he had also returned to reporting for the BBC. He was probably the most famous person alive who had been severely disabled by a bullet. It struck me that if I was to find a wounded poet, I'd find one in Frank.

The crush of excited tourists meant his approach was obscured by a line of standing figures. But he wheeled through the crowd and was apologetic for being late, shaking my hand firmly. Frank is one of those Englishmen who, in another era, would have been sent off to India to run a colonial province. He had a patrician kindness about him, with lean features and a keen mind. The way he took command of our meeting was fluid and understated – a lesson in leadership and diplomacy. He was, quite simply, charming.

We went into the building and, over coffee, he made one thing very clear. 'The BBC have been unfailingly generous,' he said. 'And the NHS have been brilliant. I quickly learned that when you are really badly shot up, with multiple injuries, you need the care of a major NHS hospital. There, some of the treatment I had was pretty pricey, like the nutrients they had to feed me through a tube in my chest to keep me alive.'

His rehabilitation treatment was extensive. And that was an important thing to say, because the treatment of the wounded in a developed nation is not cheap. One US review estimated the care costs for regular gunshots victims at $18,000.[29] This financial breakdown did not include complicated plastic or neuro-surgery, and other reviews have quoted much higher figures: $48,000 for treating people shot in the hand;[30] over $100,000 for those shot in the face.[31] Around the time Frank was shot, the daily cost of care for a spinal-cord gunshot victim in a US hospital was estimated at about $2,000 a day.[32]

It all adds up. The Pacific Institute for Research and Evaluation calculated that, in 2010, the financial burden of firearm injuries in the US came to $174 billion. They included things like work loss, medical care, mental-health fees, emergency transport provision, police time and insurance claims,[33] a bill that was estimated to cost every American $564 a year.[34] Just as the long-term pain that guns can bring is hidden, so too is the financial impact caused by them.

In some ways, if one is to see lightness where there is only dark, Frank was lucky. Sixty-two countries in the world do not have gun rehabilitation services of any kind; a shot to the spine would be the end for most.[35] In 1994 it was estimated that gun-wound rehabilitation services in developing nations reached, at most, 3 per cent of victims.[36] And today it's only about 15 per cent of people with disabilities in such developing countries who can get devices like wheelchairs.[37] The cost is huge, too – one report estimated that treating a gunshot wound in Kenya was twenty-seven times a person's average monthly salary.[38]

But these are comparative statistics and figures that offer no comfort to a Westerner who, like Frank, still has to live their days in a wheelchair and in pain. For him the agony of nerves was constant. Some days . . . and his voice trailed off, and you know he'll always be reminded that the past is real. That scar on his spine will always pull him back to a blood-soaked road and the roaring in his head.

'No,' he says, 'it is there, the pain. Like someone has kicked me hard in the shin, or shoved a screwdriver up my backside.'

And the wounded-hero fantasy I had of him was transformed, because it was clear this man of intelligence and humility, despite showing no hatred for Islam or the Saudis, can never forget what those bullets took from him. Like playing with his children on a beach. Or skiing without a second thought. Or just walking. I wanted to ask him if he could still make love. But I did not, because he was a gentleman, and I felt ashamed at wanting to know these things. He did, though, talk about feelings; he had lived through some grim moments, particularly in hospital.

'No,' he said. 'I can't say that anything good has come out of it. That would be too much.'

A shadow passed across his face.

It is easy to forget the psychological injuries sustained through gun violence, I thought. We often just associate guns with physical harm, but it is clear that's simply not the case. In one study, sixty gunshot patients admitted to a trauma centre were interviewed when at hospital, and interviewed again eight months after they were discharged. Over 80 per cent reported symptoms of post-traumatic stress disorder, or PTSD. Other studies support these findings, showing that gun-trauma patients have twice the odds of suffering PTSD than motor vehicle accident survivors.[39]

Frank had managed to find some semblance of balance again. He had the support of a major institution like the BBC, a loving family, an enviable job and a sharp mind. But he also had pain, legs that did not work, and all the secret humiliations this must bring. And his eyes were marked with it.

Perhaps that was what I took from meeting Frank. Not that being wounded by a gun brings pain; that is, perhaps, too obvious. Not that the strength of a man's character is based on whether he can turn a gunshot tragedy into triumph, because Frank saw no silver lining to what had happened. Rather, that the gun has no moral function. It does not question your worth, or your kindness, or your intelligence. It just does what it does, and that is to wound and to scar in ways that we will never truly know unless somebody shoots us – or we shoot ourselves.

4. THE SUICIDAL

The gun's bitter role in suicide – a filmed moment of despair – a dark pilgrimage to the scene of a tragedy in New York, USA – talking to an American psychologist and learning from Sylvia Plath – Switzerland – meeting a suicide charity by the shores of Lake Geneva – and an unexpected discovery

According to the World Health Organization, over 800,000 die every year from suicide, in all of its despairing forms.[1] This works out at about one person taking their life every forty seconds. What it means is that more people kill themselves each year than are killed by homicides and wars combined, and suicide is one of the leading causes of death among teenagers and adults under thirty-five.[2]

Of this mountain of dead, firearm suicides account for a huge number. Exact figures are hard to come by, but the general observation is that where guns are very common you often find a higher level of suicide deaths by that method. None more so than in the US, where more people shoot themselves than anywhere else in the world – 60 per cent of all suicides are by this method.[3] It works out, on average, at about fifty gun suicides every day.[4]

Of course, there is no such thing as one 'America' when it comes to statistics; there are major regional differences. Alaska has a firearm suicide rate 700 per cent higher than New Jersey.[5] But what we do know is that shooting yourself is becoming much more common there. The percentage of US gun suicides has increased from about

35 per cent of all suicides in the 1920s to over half today.[6] And what shocks are the quiet lines in the US data. Like the cold figures that record there were ninety-two children under the age of fourteen who shot themselves in 2011.[7]

Comparing such figures to other developed nations highlights what an unaddressed problem the US has. In England and Wales gun suicides account for less than 2 per cent of all suicides.[8] In Latin America and the Caribbean only 13 per cent of the 26,213 suicides in 2012 were with guns.[9]

Given the high level of firearm ownership there, this observation of the US as a gun suicide outlier is in line with the oft-stated link between rates of firearm ownership and suicides.[10] In fact, so strong is this link that one way the Small Arms Survey establish levels of gun ownership in a country 'is the proportion of suicides committed with firearms'.[11] So it is no surprise that the US has a firearm-suicide rate almost six times higher than most developed nations.[12]

This was why, some months before I went to South Africa and on my way back from Honduras, I had stopped over in New York. There I had travelled deep into the heart of the city to visit a place that had stayed with me ever since I had begun researching the gun's role in suicides: 1358 Washington Avenue.

The east coast wind pushed down the wide street, and clumps of fallen leaves danced in tight circles. Distant police sirens coughed out ugly staccatos across the city, and the winter sun bleached its low-covering sky. I parked my cheap rental car behind a scuffed brown Ford, a Dominican flag limply displayed in its stained rear window, and opened the door.

In front of me was a twenty-storey high, dull brown-brick tower. Bland and architecturally functional, this was one of the Projects, built in the mid 1960s to house some of the poor and the huddled masses of the greatest nation on earth. It was named after Governor

Morris, a founding father by whose hand the American Constitution was written. And it was one of ten similar blocks, housing over 3,000 people, here in the Bronx. It was not the nice side of town.

A sweatshirt and a pair of tights draped the sparse limbs of skeletal trees and flapped in the wind like flags of poverty. Coffee cups scuttled around on gritty gusts, and squirrels darted up and down the encrusted bark of the trees.

Ten years ago, Paris Lane, a troubled twenty-two-year-old, had killed himself here in the foyer of this block. Paris had imagined himself alone, but a NYPD surveillance camera was the unblinking eye that saw him talking to – and, it was later to be said, being rejected by – his sixteen-year-old girlfriend, Krystin Simmons. It captured the wiping of tears from eyes and her hugging the young man with his lank hair braids and his black, dull puffer jacket.

And it saw what happened next.

Krystin walked into the lift and the metal doors closed. Paris then, with a casualness that belied what happened next, put a balled hand into his jacket pocket and pulled out a handgun. He put the pistol in his mouth.

A life was extinguished, and the CCTV camera carried on filming.

It would have ended there, but that grainy video was to find its way onto a website called Liveleak.com, one that specialises in videos of people dying, often killing themselves. Liveleak has videos of people jumping off buildings or stepping before trains. It has a few videos of men, and it is only men, even pulling the trigger. They had posted this particular film under the line 'An Oldie but a Goodie'. So it was that over half a million people watched the ending of this man's life, and I was one of them.

The watching of a suicide played out on video is a terrible and compelling thing. You see the sudden jerk of death as the bullet rips life away, then you rewind to the point of the last breath and you pause right at that moment when life is extinguished.

These things leave digital ghosts in our mind, the grandest and most terrible of gestures. And Paris's death had haunted.

I think the seeming nonchalance with which he took his life was the thing that left its mark. The casual way he went from hugging his girlfriend to pulling out a gun. So I had travelled to this run-down district to see if I could put some meaning to his sudden end.

I walked up the path and entered the foyer. I had already been here, in a sense. It was the same, a decade on. The scuffed municipal floor, the grey metal light, the lift doors. A man in a tightly wound-down hooded top walked in. Did he know Paris? He did not answer me. Then a young African American woman came in, but she was also silent. This area had seen too much violence for people to start talking to a white guy with a notepad, so I walked back out into the wind.

School was breaking up, and, as I retraced my steps down the weed-ripped pathway, a fight broke out. The kids crowded around, screaming and jostling. A white teacher – all of the children were black – strode over, and there was a sharp bark of rebuke. The children dispersed.

An older man, solid and confident, with a flat cap and tattoos, his poodle on a leash, walked past. I asked if he remembered the shooting.

'Oh. That cat. Yeah, it was shocking,' he said. 'I had a friend who was downstairs when it happened – there was blood all over the floor. Like it was fake, you know. But the girl, the girl just carried on upstairs and didn't know. It's still being talked about.'

His name was Wayne Newton. He had lived here for forty years, ever since he was eight, and ran a barber shop down the way. It used to be crazy around here. A person like myself hanging around back then – and he pointed a finger at me and mimed pulling the trigger. I asked more about Paris.

'Yeah. You hear about him even now. "Have you seen what's on the internet? Some cat shot himself in the Projects." They don't realise he did this ten years ago.'

Paris had lived a hard life. His parents had both died of AIDS by the time he was twelve. He was an aspiring rapper who used the name Paradice. But Wayne did not know much more and so he shook my hand and told me he had a 9mm gun for himself because, hell, you needed it here once. Then his dog saw a squirrel and shivered with excitement and pulled at the leash, and Wayne was gone.

I walked the streets, around and around, the apartment block at my centre, asking if people remembered the lonely death. But Wayne was the only one who had heard of this young man, or was saying he had, at least. The mothers in the children's after-hours playgroup, each wall lined with rainbows, clicked their tongues and said they knew nothing. The police stood outside the school gates, guns on their hips and suspicion on their minds, and crossed their arms. They asked me why I was asking such questions, and I walked away from them as if guilty of something.

The woman at the housing association gave me the telephone number of a press officer on a slip of torn paper. I crunched it into a small ball and let it drop into a bin. Press officers don't talk about such things. And I walked back out into the fading evening and realised that Paris's name was long lost to these streets. His death was so famous and yet so forgotten in the very place where he died.

A slice of time can be a saving grace when it comes to thinking about suicide. One study of survivors of self-inflicted gunshot wounds in the US found 40 per cent had contemplated suicide for less than five minutes beforehand.[13]

Such a rapid shift in emotions means one thing: that the way someone chooses to end their life is important. If someone reaches for a gun, they usually don't get a second chance. If they reach for pills, they often do. Over 99 per cent of people shooting themselves in the head die, whereas only 6 per cent of people slashing their wrists or popping pills end up killing themselves.[14] And it's been shown, repeatedly, that those who survive a suicide attempt usually don't later die by suicide.

This link between the availability of means of killing yourself and the chances of a successful suicide is powerfully illustrated in the way that the American-born poet Sylvia Plath ended her life in the UK. Sylvia, despairing of the lifelong depression that had hounded her for thirty years, which she called 'owl's talons clenching my heart', killed herself in 1963. In the velvet quiet of the morning she had taken sheets of tin foil and handfuls of wetted tea towels and lined the doors and windows of her kitchen, so as to protect her sleeping children next door. Then she put her head in the oven and turned on the gas. By 4.40 a.m. Sylvia Plath had escaped the weight of her darkness.

At the time ovens in England used coal gas; something that contained carbon monoxide in a lethal dose. In the late 1950s, it was so easy to stick your head in the oven in England that nearly half of all suicides were gas deaths, some 2,500 a year. Seeing the problem, the government acted, and soon gas companies stopped using poisonous coal gas. What was known as the 'execution chamber in everyone's kitchen' was removed.

What did the suicidal do instead? It seems most carried on living. England's suicide rate dropped precipitously by one-third and then stayed at that level.[15] Many of those who gassed themselves apparently did so impulsively. As Scott Anderson wrote in a *New York Times* article: 'In a moment of deep despair or rage or sadness, they turned to what was easy and quick and deadly.' Removing the oven slowed the process down; an accessible exit from the agony of despair became suddenly less so.[16]

If this is true for gas ovens, then why not for guns? Studies have shown that just keeping a gun unloaded and storing its ammunition in a different room significantly reduces the odds of that gun being used in a suicide; it seems that there was some correlation between access and lethal action. So I contacted one of America's leading experts on suicide – Paul Appelbaum, a professor of psychiatry at Columbia University – to find out more. We spoke over a crackling line, but what he said was clear.

'There is probably no psychiatrist who couldn't tell you they know patients who have thought about ending their lives,' he said, his voice dipping in and out on the call, 'who then, with the treatment for depression or alcoholism or whatever, went on to live thirty or forty years without a recurrence of that kind of suicidal ideation. Suicidal ideation is often situation-specific and often transient.'

He talked about the people who throw themselves off the Golden

Gate Bridge in San Francisco – one of the world's most popular sites for suicides. The bridge stands 75 metres above the cold water of the Frisco bay. After a fall of four seconds, jumpers hit the surging surface at 75 miles an hour, a speed that usually breaks their backs. Over 95 per cent of the jumpers die from the drop alone, the rest from drowning or hypothermia. The chances of dying are about as sure as using a gun, so the bridge attracts those who see a flying death being better than a blood-splattered one, with a peak of ten suicides there in one month alone.

'Years ago, back in the late '70s, a professor at Berkeley collected information regarding over 500 people who attempted to jump from the Golden Gate Bridge but were restrained,' Paul went on. 'He found the vast majority, almost 90 per cent of them, were still alive twenty-five years later, and only about 5 per cent had subsequently committed suicide, 5 per cent dying of natural causes. OK, the rate of suicide was higher than the general population, but 90 per cent of them were still alive an average of a quarter of a century later.'

His words made me think about Paris. Was it inevitable he was going to kill himself? If he had not been able to get a gun, would he have wandered down to the dark, slick spread of the Hudson River and drowned himself instead? Or would he have waited for the next lift to the top floor and thrown himself into the yielding air and felt, for a brief second, like he was flying?

These were questions that I had sought an answer to, not in the US, but back when I was in Switzerland visiting the Small Arms Survey. That was because there I had found other data worth exploring. Not about the numbers of guns, but about the numbers of suicides by them.

After leaving the Small Arms Survey, I had walked back out into Geneva and followed the tramlines south, passing the quiet classical façades that line its neat streets and from there crossed the slow-pushing waters of the Rhône onto the Boulevard Georges-Favon. The sky was a peppered grey, and genteel apartments rose on each side. I carried on until I reached a place where the air was filled with the chaos of crossing cable-car wires and the roads filled with bookshops and Chinese tea merchants.

A sun-frayed shop window caught my eye; a Dungeons and Dragons store. Fantasy and comics do well here, because the Swiss seemingly enjoy such subtle distractions along with chocolates and fancy timepieces. There, in the window, a yellowing box stood: 'Descente: Voyage dans les ténèbres'. It was a game that led you down into the shadow-lands and beyond. It seemed apt.

Around the corner, opposite a café selling bitter coffees and neatly stacked pastries, was my destination: Stop Suicide. It was a charity set up to help young people at risk. Sophie Lochet, a conscientious woman in her mid twenties, met me at the door wearing a white shirt, blue jeans and a pair of glasses that framed a kind face. She beckoned me into a room filled with reports and posters, campaign leaflets and books with depressing titles. I settled down amid offers of coffee and chocolate.

'In Switzerland, every day, about four people will take their lives,' she said. Over a thousand a year. 'That's three times more than die in road accidents.' She explained the role that guns play in this.

Switzerland has one of the highest number of guns per household in the world. This small landlocked state, high on the west-central plains of Europe, has almost 3.5 million firearms in a population of just 8 million, leaving almost 40 per cent of Swiss households with one.[17] This, she said, had led to guns being one of the leading ways of suicide among young men caught between the volatile ages of fifteen and twenty-nine.[18]

One reason for this, Sophie explained, was that – until recently – Swiss conscripted soldiers were allowed to keep their rifles with them after they had completed their military duties. About 40 per cent of gun suicides here were at the end of an army-issued weapon. In 2003, though, the number of Swiss soldiers was halved as a result of a sweeping army reform. This sudden decline in the armed forces meant there was a parallel decrease in the availability of guns nationwide.

This was why I had come to see Sophie. I wanted to understand if this drop in guns had resulted in fewer suicides, to see if those who say, 'There is no correlation between gun control laws and murder or suicide rates,' were right.[19] I wanted to challenge what the pro-gun lobbyists would argue: that 'denying one particular means to people who are motivated to commit suicide . . . simply pushes them to some other means'.[20]

Sophie had the answer. 'One study did look at the gun suicide numbers before and after that reform. The academics found there was a major reduction, both in the overall suicide rate and also in the firearm suicide rate.' The evidence, she said, was compelling. Echoing what Paul Appelbaum had told me, only 22 per cent of the reduction in firearm suicides was substituted by other ways of killing oneself. Like the coal gas situation in the UK, a drop in access to guns in Switzerland was not followed by a rise in other forms of suicide. Rather, it led to an enduring drop in the general suicide rate. Today in Switzerland there are about 200 gun suicides per year. Two decades ago it was about 400.[21]

Switzerland is not the only place this sort of cause and effect has been seen. I later read that in 2006 the Israeli Defence Force witnessed a disturbing number of suicides in its ranks. In an effort to bring down this number, the IDF banned soldiers from taking rifles home on the weekends. Suicides fell by 40 per cent. An army review concluded: 'decreasing access to firearms significantly decreases rates of suicide among adolescents'.[22]

In Australia, too, a series of strict gun control laws in the mid 1990s led to a decrease in gun suicides, but with no significant rise in other types of suicides.[23] These findings are in line with a host of studies that have found, again and again, that stricter gun laws are associated with lower gun suicide rates.[24]

Yet, in the US, where about 20,000 people die every year from suicide by guns, the political classes seem inured to such observations, and pro-gun advocates deny any correlation between access to guns and suicides. The lobbyists, most notably the National Rifle Association, conclude fatalistically that gun owners 'exhibit a willingness to take definitive action when they believe it to be in their own self-interest.

Such action may include ending their own life when the time is deemed appropriate.'[25] They do not take into account the influence of mental illness on gun suicide.

But Sophie was convinced. Guns cause suicides. Of course, other sorrows play their part. Suicides are driven by depression, loneliness and broken hearts, but guns help transform a moment of crisis into a final act.

And yet firearm suicides somehow are still seen as inevitable and unstoppable. Or they are called something else entirely – an accident, a gun discharge. I had read that in some countries, gun suicides are under-reported by as much as 100 per cent.[26] It was like that with Ernest Hemingway, a man long beset by depression. He had 'accidentally killed himself while cleaning a gun this morning', according to his wife.[27] But we know what really happened. We know even without the *New York Times* pointing out that Hemingway was 'an expert on firearms', or that his father had taken his own life with a Civil War pistol.

Such reluctance to admit that suicide occurs, to address the things that contribute to it, such as gun availability, stems, it seemed to me, from the idea that taking your own life goes against nature and God, a view long held in Anglo-Saxon society.

'Self-murder' became a crime in England in the mid thirteenth century, while the term 'to commit suicide' reflected the Catholic Church's view of the act as sinful. Suicide victims were once denied a Christian burial, dragged to a crossroads under the cloak of night and there thrown in a pit, a wooden stake hammered through their hearts. There were no clergy, choir or prayers. The family was even stripped of their belongings, which were given to the Crown. In this way, the suicide of an adult male once could have reduced his survivors to poverty.[28]

Even today, in some parts of the world, suicide is still illegal. In India, you could, until 2014, have faced up to one-year imprisonment for trying to shoot yourself. You can still be sent to prison for attempting to kill yourself in Ghana, Singapore and Uganda. Internationally, we still have a social disgust for suicide as well as an institutional refusal to respond to it. No wonder it is rarely talked about when it comes to gun control.

I thought of this as I watched the video of Paris Lane's death once again. Half a million people might watch his death, but nothing changes. No campaign is launched, and sympathy remains muted. Such is the power of the gun and the silent powerlessness of those in despair. The gun not only transforms these people's decision in taking their lives, but it also transforms our response to their deaths.

The comments below Paris's video revealed this: 'That man was weak and let his infatuation for a woman overpower his rationale,' wrote one.

'The hell is waiting you idiot,' added another.

Some were just ugly: 'If all blacks did the same America will be a safe place to live.'

But then I saw a comment that surprised me.

Someone had written: 'They didn't break up u fools . . . He had problems with some bad guys and they hunted him. He went home to his mom and gf to say goodbye then killed himself so they wont kill his family and friends . . .'

In reading this, my view of Paris Lane's suicide changed. Perhaps his death was more a response to power than to pain, I thought. And in so thinking, the gun took me down a path that led away from those impacted by guns, towards those who wield them.

III. Power

5. THE KILLERS

The world of mass shooters and assassins – Finland recalled – a bloody day in a teaching college – the American mass shooter examined – Norway – travelling into the wilds – the scene of the worst mass shooting in history – an Oslo drink with a killer's expert – a meeting with Julian Assange in London – the ugly offerings of the dark web

There are many types of people who kill other people with guns.

There are those who do so for the explicit control of power. They are, by and large, criminals, gang members, terrorists, policemen or military personnel. When they pull a trigger and a life is ended, people in one of these groups do so out of obedience to a specific ideology or dogma. It could be a desire to control the streets, an urge to rob, to protect their nation, to maintain order, even to exact retribution. Of course, when they take a life they all become killers. But their actions are generally not driven by a desire to kill for killing's sake. Death is a by-product of something else – usually power and control.

There are the untold numbers of killers who are gun owners who use their weapons for acts of personal power. Those caught in a moment of passion, despair, anger or self-defence, who use their guns to take a life. Sometimes their actions are justified, often not. These deaths are usually not premeditated; rather they are a specific response to threats, passions or fears. And the motivations behind this use of the gun are so diverse that to understand what makes such people kill is as complex as understanding life itself.

Then there are those two groups who seek a darker form of power – for whom death is the thing they seek. Killing not as a by-product of protection, or defence, or desire, but death as a means to its own powerful end. These are the mass shooters: all too often young men who go on the rampage and kill in a single, public event. They defy the normal motives for violence – robbery, envy, personal grievance. They ignore basic ideas of justice.

Or the assassins. That rare breed of cruel men who are paid to kill, and for whom money cannot be the only reason they do this job, because you can always earn a living doing something else.

Mass shooters and assassins: two groups I felt had to be the first things to write about when I shifted my gaze away from those whose lives were lived and ended at the end of a gun and entered the world of those holding the gun in their hands.

The snow was beginning to fall on that September day in 2008 when I walked through a muffled forest that surrounded a small town in western Finland. I had travelled out to the creeping edge of Kauhajoki and there, caught within an endless line of trees, was trying to find a rifle range. A place where a now-dead man had been filmed a few days before, shooting at targets, and where his poisonous words foretold a horror.

I had left the road ten minutes earlier and cut into the claustrophobic woods and was now lost. The sound of my feet breaking the frozen ground cut the quiet. And then, ahead, between the pines, I saw the outline of a wooden shooting range.

The cracking of the leaves startled the man. I first noticed him as he began to turn; his coat blended well with the leafy surrounds. Then I saw the rifle loosely cradled in his arm. This was not the time, nor the place, to meet a stranger with a gun.

A few days before, a young Finnish man, a twenty-two-year-old called Matti Juhani Saari, had done the unthinkable. He had walked into his college and killed ten people. Saari had gone on the rampage about 5 miles from here, at the Kauhajoki School of Hospitality. He had crept into the university buildings through the basement and, armed with a Walther P22 semi-automatic pistol, wearing a balaclava and dressed in military black fatigues, had gone upstairs. He acted as if he was on a combat mission, but he was the only enemy in this quiet Finnish town.

At 10.30 that morning Saari had walked into an exam room, filled with his fellow students taking a business studies paper, and opened fire. He approached his victims one by one, shooting each at close range. He then stepped into the corridor, loaded a new clip and returned to kill his teacher. He slowly moved around the classroom, delivering a vicious *coup de grâce* to whoever made a sound.

After he had killed, Saari called a friend to boast about what he was doing. He then poured petrol onto the crimson-stained floor, dropped a match and walked outside, the fire rising behind him. As the flames rose, nine classmates and one teacher lay dead and eleven more injured in the cruel flickering. Saari watched the rest of the students running, screaming, out into the thin light of a Finnish autumn, then he shot himself in the head.

With a total of eleven people now dead, it was the deadliest peacetime attack in Finnish history. Saari had fired 157 shots, sixty-two of which later were found in the bodies of his victims.[1] Twenty rounds were in one person alone.

One bullet, the least lamented, was the round he used on himself.

With that final shot his pistol also sounded the start of a different race, a race for journalists to get to the scene, to report on the terror created by that most modern of things: the mass shooter.[2]

I happened to be in Oslo at the time and, because news desks in London do not think: 'It would take him seventeen hours to drive there, and it's 700 miles away,' but rather: 'Norway is close to Finland, so let's send him,' they called me.

'Get packing. We're going to send you to Finland.'

And that was it. I was with Jenny Kleeman, an up-and-coming journalist, and we were reporting for ITN on Norway's immense oil wealth. We were analysing Oslo's sovereign investment funds when we got the call, and death was the last thing on our minds. But a day later we had flown (not driven) to Kauhajoki, a place forever marked by what had unfolded there and one that left its mark on me – because it was my first encounter with the grotesque realities of mass shootings.

We hit the ground running. My editor back in London was hungry for facts about why Saari had done what he had done, and we quickly learned that the troubled killer, in the weeks leading up to the incident, had posted several videos online under the username 'Wumpscut86'. His terrible message: 'You will die next.' The videos showed him firing his Walther P22 at a local range.

That was why I was lost in that wood. I was at that range – Saari's last location captured on film. The young killer was dead, but at that moment no one knew who had been behind the camera – an accomplice, even? I looked down at the man's rifle, and my mind clouded with the possibilities.

The man turned and stared intently. Then he tutted. And I realised what I had first mistaken for rage was actually annoyance. Saari had just filmed himself, and this man was not going to kill me. He was just bothered that I was here, stumbling about in the forest with my video camera. Because my presence, in this remote province of this little-visited country, was a clear signal to him of what was to come: a bloody media spectacle.

The modern mass shooter and the modern media are intrinsically linked. Columbine, Dunblane, Sandy Hook: journalists, responding to the final performance of a lone shooter, have ensured that these place names are forever marked. In news 'if it bleeds, it leads', so the saying goes, and that evening the news the world over led with the blood Saari had shed and the name of Kauhajoki. Bulletins showed the rows of flickering candles and teddy bears outside the school. Images of the Finnish emergency services standing around awkwardly were transmitted across the world. And the shooter's vicious videos and his ugly testimonies were replayed endlessly.

Of course, this was a big story, not only because it was the second mass shooting that Finland had had in two years, but also because it was the death of young, hopeful white people. Such things are important in Western news agendas, because prejudices and priorities dictate the amount of airtime a story is given – what has been called a 'hierarchical news structure on death'.[3] A white shooter killing twenty kids in the US will dominate the global press. Twenty black adults dying in a hail of bullets in Nigeria will barely register. And when it comes to mass shootings, schools will always get more coverage than anywhere else, even though in the US businesses are almost twice as likely to be the blood-soaked epicentres of mass shootings.[4]

What it means is that, while mass shootings may only constitute about 1 per cent of all gun deaths in the US, their impact in terms of headlines and column inches is profound.

Some say it is too much, that the media's saturated coverage of a mass shooting encourages others to carry out copycat attacks – tortured souls seeking to burn out in a blaze of infamy.[5] They have a point. In 356 BC, a Greek called Herostratus torched the Temple of Artemis at Ephesus. It was, contemporaries wrote, an attempt to immortalise his name.[6] And it worked. The fact we know the arsonist's name, the destroyer of one of the Seven Wonders of the Ancient World, shows us that terrible crimes can achieve eternal fame. In the same way we know the names of Adam Lanza, Seung-Hui Cho, Anders Behring Breivik and, possibly in small part because of my efforts, Matti Saari.

This idea that the media can influence extreme behaviour is perhaps best illustrated by looking at the time when newspapers agreed to cooperate with the authorities following a spate of suicides in the 1980s subway system in Vienna. Negotiations led to local Austrian papers changing their coverage by avoiding any simple explainers as to why someone threw themselves in front of a train, by moving the tragic stories off the front page and keeping the word 'suicide' out of headlines. Subway suicides there fell by 80 per cent.[7]

This led some to ask: 'Would the same happen if there was a media blackout on mass shootings?' Certainly there have been very vocal critics of the media's saturation coverage of some mass shootings. A forensic psychiatrist told ABC News the airing of the Virginia Tech killer's video tape was a social catastrophe: 'This is a PR tape of him trying to turn himself into a Quentin Tarantino character

. . . There's nothing to learn from this except giving it validation.'[8] Others have said that the gory details of shootings help 'troubled minds turn abstract frustrations into concrete fantasies'.[9]

Perhaps these things are true. But the media's focus also highlights things like the inadequacies of existing national gun law. The fierce coverage of Kauhajoki, for instance, encouraged the Finnish government to reduce the number of handgun licences and to raise the minimum age of gun ownership. The media helped do that.

So, when journalists descend on a sleeping town where lives have been shattered by the sharp retort of gunfire, they should tell themselves they are there to report on these horrors for one reason and one reason only – to try to stop this happening again. Not to titillate, but to warn.

The reporter and I thought of these sensitivities as we lined up outside the school's entrance that night – a straight run of white broadcast trucks in front of pools of candles and stunned locals. Then London called, and we were on-air.

In 1966, a twenty-five-year-old ex-marine called Charles Whitman climbed to the top of the University of Texas tower. He carried with him three rifles, three handguns and a sawn-off shotgun. By the time he was killed a few hours later, Whitman had shot forty-eight people, sixteen of whom died, and the world was introduced to a very unique, modern monster: the mass shooter.

Of course, the terrible visitation of mass death on schools and offices is not just an American tragedy. The deadliest mass shooting was by Anders Behring Breivik in Norway in 2011, where sixty-nine died in a shooting spree, and a further eight lost their lives in a bomb blast. Before that, the world's deadliest attack by a lone shooter was in a small farming community in South Korea. There, in 1982, a policeman called Woo Bum-kon killed fifty-six. His rampage was triggered when the woman he was living with woke him from a nap; she had swatted a fly that had landed on his chest.[10]

Despite these global killings, the greatest media focus is on those carried out in the US. An Associated Press list of twenty of some of the 'deadliest mass shootings around the world' featured eleven US attacks.[11] It's been calculated that there have been over 200 such incidents in the US since 2006.[12]

And if you define a mass shooting as one where at least four people are wounded, not killed, then in 2013 there were 365 American incidents: a mass non-lethal shooting every single day.[13] It's also seemingly getting worse. According to the FBI, the rate of deadly mass shootings went up from one every other month between 2000 and 2008 (about five a year) to over one per month between 2009 and 2012 (almost sixteen a year).[14] Of the dozen deadliest shootings ever to have taken place in the US, half have been since 2007.[15]

Of course, the media focus not just on the numbers killed and the frequency of the killings, but also on the people who wielded the guns. People ask: 'Who would do such a thing?'

It's difficult to give a definitive answer. The US secret service looked at the phenomenon of mass shooters and concluded there was no single 'profile' of a school shooter. Each shooter differed from others in numerous ways. Despite this, there is a consensus that some trends exist. In 2001 a study looked at forty-one adolescent American mass murderers: 34 per cent were described as loners; 44 per cent had a preoccupation with weapons; and 71 per cent had been bullied.[16] Other traits seem to dominate, too.

Mass shooters are almost always male. There's only been a handful of cases of female mass shooters: one such was Jennifer San Marco, a former postal worker, who killed five at a mail-processing plant in California, as well as her one-time neighbour, before shooting herself.[17] Why mass shooters are so disproportionately male is unclear. Some see men as having a different approach to responding to life's disappointments. Others see their violence as highlighting gender differences in testosterone levels and mental development.[18] Each reason is frustratingly nebulous, though, and, apart from banning access to guns to all men, does little to help us work out how to put an end to such murders.

Mass shooters are loners. In rare instances, there may be two shooters working together, such as in the Jonesboro massacre, where Mitchell Johnson, aged thirteen, and Andrew Golden, just eleven, shot dead four students and a teacher and wounded ten others.[19] But, generally speaking, a mass shooter typically acts alone and is not affiliated to any group or cult, again making it hard for authorities to identify them and act preemptively.

They are relatively young; the Congressional Research Service puts US mass shooters average age at thirty-three.[20] It's rare for them to be very young, though – ages eleven and thirteen are untypical. There are various things that go towards explaining why adolescents don't go on rampages: children's access to guns, the fact that teachers and parents are often able to intervene when adolescents exhibit worrying behaviour, and the reality that shorter lives are often not so filled with disappointment all play a part.

We know that mass shooters are typically socially awkward. They rarely have close friends and almost never have had an intimate relationship, although they sometimes have had failed flings. They don't tend to have problems with alcohol and drugs, and they are not impulsive – indeed quite the reverse.

This might lead many to assume that mass shooters are all blighted with a long history of mental ill health. Not so. Obviously they all have a warped and broken view of the world to do what they do, but a diagnosed mental-health condition is an extremely poor predictive factor for profiling whether someone is likely to go on to become a mass shooter. A 2001 analysis of thirty-four American mass shooters found that only 23 per cent had a recorded history of psychiatric illness.[21]

Despite this, we still fixate on the mental oddities of these troubled men. We comment on the fact Martin Bryant, who carried out the Port Arthur massacre in Australia, was really into the soundtrack of the *Lion King*.[22] We write how Adam Lanza, the young man who murdered so many children at Sandy Hook, carried a black briefcase with him, while other students had backpacks. We recall how Seung-Hui Cho, the warped killer of thirty-two at Virginia Tech, enjoyed taking photographs up the skirts of fellow students under the desks with his cell phone.[23] But these are traits that, while odd, are far from proof of a mass murderer in the making. As one psychologist put it: 'Although mass murderers often do exhibit bizarre behavior, most people who exhibit bizarre behavior do not commit mass murder.'[24]

Nonetheless, it is fair to say that mass shooters are often very focused outsiders who plan their actions obsessively. Many massacres have been in the pipeline for months, sometimes years: the Columbine shooting took thirteen months to plan.[25] Anders Behring Breivik in Norway claimed he had been plotting his actions for five years.

This planning reflects a fixated and resentful view of the world. Mass shooters want to fix their ideas in history: a sort of personal vindication through gunfire. Whereas terrorists use guns and the media to promote political or religious beliefs, mass shooters use guns and the media to highlight their own personal grievances. Like Virginia Tech gunman Seung-Hui Cho, who sent NBC News an 1,800-word statement and twenty-seven QuickTime videos with him ranting to the camera.[26]

Other trends emerge. Many mass shooters take their own lives.[27] Many wear tactical military clothing. They often use high-powered and rapid-fire weapons. The guns used in sixty-two mass shootings over the last three decades were looked at by the website Mother Jones. Over half involved 'semi-automatic rifles, guns with military features, and handguns using magazines with more than 10 rounds'.[28] One of the guns that James Eagan Holmes used to shoot seventy-one, killing twelve, in Aurora, for instance, was a semi-automatic rifle with a 100-round drum magazine.[29]

This use of such legal weaponry should concern. FBI data shows that, between 2009 and 2012, mass shootings that involved assault rifles or high-capacity magazines led to an average of sixteen people being shot, 123 per cent more than when other weapons were used.[30]

These are ugly and disturbing observations and statistics. But they only served to help a little in my analysis of the world of the lone mass shooter. So I looked at the long list of perpetrators again, seeking someone who was, perhaps, representative of all of these trends.

I was searching for an archetype – a shooter who was relatively young, alone and socially awkward; someone who wore a uniform and carried a semi-automatic rifle with high-capacity magazines; someone who was not clinically insane; a fantasist who had penned an angry manifesto. This Venn diagram of horrors showed up one ugly and familiar name. The most murderous mass shooter of them all: Anders Behring Breivik, the Norwegian right-wing killer.

When I arrived in Norway to learn more about Breivik, the only car the Oslo hire company was able to lease me ran on electricity. I had never driven an electric car before and, as I headed through the dynamite-blasted grey mountains that led from the capital, I was disturbed to see the number on its power-gauge drop dramatically. It had read 123 kilometres when I pulled out of the car park. Now, 50 kilometres out, the power meter read 13, and I had 18 more kilometres to go. The beginnings of mild panic shifted up my spine, because the cold here was profound, and there were few car-charging points. Images of freezing to death in a Norwegian electric car – a hypothermic victim to a green response to global warming – dominated my thoughts.

As the power dipped, so did the sun, casting its last shallow, anaemic light across the deep and broad lake of Tyrifjorden. Beyond lay Utøya: the outermost island. Its name was still hard for some to say because this was where Breivik had killed dozens.

Then, as the gauge told me I had two kilometres of power left, the Sundvolden Hotel, one of the oldest inns in Norway, came into view. Framed beneath the pine-rimmed peaks of King's View and the stretching cold-blue lake, it had a beauty unique to Scandinavia.

Its Gildehuset, with its tenth-century metre-thick walls and its foyer lined with the statues of glass-eyed bulbous trolls, could be considered idyllic. But this place will not be remembered for Norwegian fairy tales or Viking walls. It will be forever marked by what happened in 2011, because this is where the survivors staggered from the worst mass shooting by a single gunman ever recorded. And these rooms were filled with grief-wrapped relatives waiting for the cauterising news of how their sons and daughters had died.

A few days before the shooting, about 600 people, mostly between fourteen and twenty-five years old, had gathered on the pine-lined island of Utøya, across the lake, for their annual summer camp. They were diverse and liberal – the cream of Norway's Labour Party youth. But Anders Behring Breivik, a thirty-two-year-old from Oslo, saw betrayal in their tolerance and weakness in their ideals. So, on 22 July, he took a boat over to the island, hollow-point bullets filling his pockets.

Breivik shot his first victim just after 5.20 p.m. He gave himself up to police seventy-five minutes later, and by then sixty-nine people had died. He had fired 297 shots – 176 with his Ruger and 121 with the Glock. Eight more people were killed, and over 200 injured, by a fertiliser bomb that Breivik had detonated in Oslo's government district an hour and a half before he began his island rampage.

He carried out these horrors wearing the uniform of a police officer, playing on a trust in the state that is implicit in so much of Norwegian life. He was also wearing earplugs to protect himself from the sound of his gunshots. Two ugly details that tell you much about the man.

He was indiscriminate and brutal in his killings. He usually fired only when he was certain to hit, killing slowly and methodically with headshots at very close range. He said to those hiding in the bushes 'Don't be shy', before he shot them. Others he murdered as they held on to each other. Stuck on the suddenly claustrophobic island, some students braved the freezing waters and swam to safety. They were plucked like white gulls, bloodied and blue, from the hard rocks of the shore.

Of the sixty-nine dead, sixty-seven had died from being shot, one drowned and one fell to their death from a cliff. Thirty-three of them were under eighteen years old. The youngest victim, Sharidyn Svebakk-Bøhn of Drammen, was just fourteen.

I fell asleep in my ancient bedroom with that thought.

The next day, my car recharged, I drove back out into the March light, the King's View behind me, forests of pines and snow beyond. The silent road ran 30 metres above the shore, and Utøya stood beyond, distant and inaccessible. There were two signs pointing the way to the island but no bridge. Precisely why Breivik chose it for his massacre.

Where the route led down to the jetty someone had put up three plastic chairs and a sign that read 'Private'. Beside it stood a row of neat postal boxes in wood, each hand-painted. The names on them spoke of long lineages and deep histories: Johnsrund, Aamaas, Syverson. Behind, on a stone, stood five memorial candles, gutted and unlit, circling a wet and dirty teddy bear. The Norwegian flag lay limply to one side, and a pine tree stood covered with broken decorations on its dripping leaves. A Christmas not celebrated.

I drove further along the road to a campsite and pulled up near the main house. The bone-marrow chill had forced me to wear all the clothes I had. I eased out of the car and waddled over to ring the bell. Nothing. But as I headed back, a man in heavy blue fatigues, black cap and thick boots came towards me through an icy drizzle that had punctured the morning. His name was Brede Johbraaten, the owner of the camping ground. I asked him about renting a boat to cross to the island, but he said this was still winter, and people did not hire out boats in winter.

He would answer some questions, though, so we sheltered from the growing rain in a wooden workshop. This man in his mid sixties, a grandfather of three, was quiet at first, but then he began to speak about how he had helped people, dripping and terrified, out of the lake on that terrible day. He had run this campsite since the 1990s, with regular visitors from Norway, Germany and Holland, but the shooting had deeply hurt his business.

'I'm fed up with it,' he said – a Norwegian understatement.

As he spoke he became more critical. First, he blamed the police, as people often do when tragedy arrives unbidden, because we need to blame someone. He said they had been too slow to respond, too disorganised. But so rare are mass shootings in Norway that you could understand why there was such confusion.

Then he said journalists come here and all they want to talk about is what happened on that day, and not what had happened to the community. So I began to ask him questions about his life, but I floundered. What had happened here made me feel almost shy. I was hesitant to talk about the horrors that had unfolded. So I asked if house prices had been hit, and we talked a little about this, as it was something we both understood.

Then, as if he felt obliged to, he spoke about Breivik.

'He is a stupid man. They should not call him by his name. They should call him as the mass killer, and that is that.' The easiest thing would have been if someone had just put a bullet in the head of that mass killer, he said.

The rain had begun to fall harder now, and the lake sparkled with the drops. There was not much more to talk about, or those things that could be said felt wrong to say. So we shook hands, and I left him, the permanent view of the island framing his land, and I wondered what it must be like to wake every day being reminded of what happened here.

Further along the lakeshore I parked at a small strip of rock that projected out towards the island. Here the government intended to set up a permanent memorial, a sharp cut-through grey stone to symbolise the unnatural tragedy that had engulfed this place. I sat and, through my misting windscreen, watched as the white clouds slid down from the mountains and shrouded Utøya.

I had been to a few places around the world which had been marked by guns, just as Breivik's guns had done here. School massacres in Britain and America, mass graves in Somalia and the Philippines, genocide sites in Armenia and Germany. That same awkward quietude, that feeling that any question you ask is tinged and mawkish, an absence of any easy explanation for what happened – these things were always a feature. So it was here. And the space between the earth and the heavens grew slowly smaller as the clouds came in, and the rain lessened until silence was the only thing left.

Night had returned to Oslo's glistening streets as I walked past endless shops selling kitchens and homeware: white candles and wooden floorboards and Scandi-chic. Norway does not wear its wealth loudly. Things here are not grotesque or baroque. But good taste requires consensus; social order and criticism are there in case you step out of line.

But where Norwegians see good taste and a proper way of living, others see intolerance and small-mindedness. Because, beneath the liberal attitudes, a provincial conservatism lurks. Norwegians might be friendly, open-minded, polite even, but you can't escape the impression that some think they are better than you.

This, at least, was what a Pakistani taxi driver, who had once been a PhD student in Islamabad, told me in Oslo. He had grown a beard since they had taken his taxi licence photograph – he had rediscovered his Islamic faith in the Fjords. He spoke about the perpetual unsaid: that if you don't like the rules of Norway, you had better go back from where you came. But it's hard to say no to living in a place with one of the highest qualities of life in the world.

I thought about the driver's words as I walked the neat streets and wondered perhaps where he saw intolerance, if others would see just a strong sense of conviction. You need self-assurance to have good taste and a highly functioning society. But with light always comes dark, and it was this national self-belief that, perhaps, found its most aggressive, most self-deluded form in the mind and actions of Anders Behring Breivik.

Such reflections occupied me, because I was on my way to meet a Norwegian writer, Aage Borchgrevink. Aage had spent many months investigating Breivik and the motivations that drove him to kill, and I wanted to know if such a murderous gunman as Breivik can operate outside the culture he wants to annihilate.

Aage was handsome without vanity. Wearing a high-collared grey sweater and a blue T-shirt, he was the sort of person you'd cast as a good guy in a Scandinavian police series. His English was impeccable. But he was, in a way, not a typical Norwegian. He had been a human rights investigator in the Balkans for over twenty years – Chechnya, Belarus, the Caucasus. He was self-critical and had lived long enough outside Norway to see it for its flaws as well as its beauty.

We met in a bar called Den Gamle Major, the Old Major, a place where Breivik himself was likely to have once drunk. I walked up to the counter and bought a glass of wine for Aage, a beer for me. It cost $30, and I had to ask twice to make sure I had heard the price right. But it was right, because Norway has the second-highest alcohol taxes in the world: the price of social order contained.

Taking the two glasses back to the table, Aage was quick to get to the matter at hand. We began at the beginning, as you do with such things: with the killer's relationship with his mother.

Aage explained that Breivik's family problems were well docu-mented by mental-health workers. When Breivik was just four, his mother became preoccupied with the fear her son would violently assault someone and frequently told him she wished he would die. Psychiatrists in the 1980s had concluded that the timid boy was a 'victim of his mother's projections of paranoid aggressive and sexual-ized fear of men in general'.

Despite these terrible reports, Aage said that the response of the state was not to intervene. Their reaction to this abuse was moulded by a strong Norwegian self-conviction about what was right and wrong. In this case, Aage said a belief in biological determinism – that the ideal condition for a child was always to be with their mother – was presumptive at the time. Both the court and the Child Welfare unit disregarded the warnings of experts. Breivik stayed with his mother and his father was not granted custody.

'The system,' Aage said, 'let him go.'

Such a failure to intervene meant, Aage thought, that there was a missed chance to stop the young boy evolving into the deeply troubled young man. Under cross-examination, even Breivik said his mother was his 'Achilles heel': 'the only one who can make me emotionally unstable'. The killer told the court he would urge his mother, a solitary woman, to find a hobby. She would tell him, 'But you're my hobby.'

Then there was the sexual nature of their relationship.[31] Aage said that social workers put in their reports 'the mother and Anders sleep in the same bed at night with very close bodily contact', but nothing was done about this. As an older man, Breivik would sit on top of her on the sofa and attempt to kiss her. He even once bought his mother a dildo.[32]

The psychological impact of this childhood clearly distorted Breivik's view of the world and of himself. 'He was almost like a zombie,' Aage said. 'His manifesto was very consumer-driven but it was lifeless. He defined himself by his brands. He'd go and buy a sushi dinner for a hundred dollars or go and buy a thousand-dollar outfit. It was a form of hyper-consumerism.' It was a lifestyle funded by Breivik's selling of fake diplomas online – a scam from which he made millions of Norwegian krone.

The interesting thing, though, was how much the Norwegian legal debate during the killer's trial appeared to focus on the psychological past of Breivik. There were two forensic psychiatric reports done on him. The first came back with the diagnosis that he suffered from paranoid schizophrenia – making him criminally insane. The other was a diagnosis of a compound personality disorder, with an emphasis on narcissism and paranoia – meaning he was criminally sane. The court settled for the second view.

The media, in turn, fixated on issues like right-wing extremism in Europe, the ability of the internet to help radicalise young men and the failures of the police on that dark day for not pre-empting the attack. But one issue was largely ignored in all of this.

'No, there was not much debate about gun laws,' Aage said. This surprised me. In the US most massacres stimulate the gun law debate. But here in Norway it was the focus on society and on Breivik's upbringing that dominated.

Even by Breivik's own account, though, guns and military paraphernalia were central to his planning. He had to overcome a problem – in Norway it's not that easy to get your hands on a gun.[33] So he spent six days in Prague in the early autumn of 2010, because he believed the Czech Republic's gun laws were amongst the most relaxed in Europe and that he would be able to buy what he wanted there: namely, a Glock pistol, hand-grenades and a rocket-propelled grenade.

Before Breivik left Norway, he even hollowed out the back seat of his Hyundai to clear space for the firearms he intended to buy. But he failed to get any, writing later that Prague was 'far from an ideal city to buy guns'. His only 'success' was having sex twice.

Returning to Oslo, Breivik ended up buying his weapons through legal channels. He said in his manifesto he could do this because he had a 'clean criminal record, hunting licence, and two guns already for seven years'. In 2010 he got a permit for one more gun: a $2000 .223-calibre Ruger Mini-14 semi-automatic carbine; he said he was buying it to shoot deer.

The next thing he wanted was a pistol, but getting a permit for that proved much more difficult. He had to demonstrate regular attendance at a sport-shooting club and so, from November 2010 to January 2011, Breivik went through fifteen training sessions at the Oslo Pistol Club. And with each lesson, his ugly plan came closer and closer to its bitter end, like a spider patiently waiting for a killing.

By mid January his application to purchase a Glock pistol was approved. He then bought ten thirty-round magazines for the rifle from a US supplier, and six magazines for the pistol in Norway.

The rest we know.

Perhaps because it took Breivik so much time to arm himself, or perhaps because of a wider refusal to believe that firearms had a pivotal role in the massacre, guns did not play a major part in the debates following the killings. There was a brief suspension of Norwegians' ability to buy semi-automatic rifles, but the hunting lobby there appears to have influenced the policy-makers, and that law was quietly dropped.[34] And today Norway still allows semi-automatic guns.

In a country where reasonable debate seems so lauded, this struck me as odd. Clearly, the numbers that Breivik killed was partly down to his having trapped the students on an island. But the fact he could shoot and shoot again without having to pause to cock his rifle must have given the children he was shooting less time to run into the trees and hide.

Rather, it seemed that for a society like Norway, one that has such self-belief, to comprehend what Breivik had done they had to focus more on the failure of the individual, his mother and the police response, not on their own gun laws or their own failings as a country. Maybe this was the right response, though. After all, you can't let one idiot with a gun change the way you live. If you do that, then they win.

With that thought, I said goodbye to Aage. The blackness of what the mass shooter was capable of was in danger of consuming my attention. The more you looked into that abyss, the more your gaze was held. So I shifted my focus onto something else, but equally sinister: the way of the assassin.

Perhaps it should come as no surprise that the person who first inadvertently introduced me to the dark world of the assassin was wearing a bulletproof vest at the time, precisely because he feared one of their bullets.

It was a late springtime London, 2010, when I received a phone call from someone I had never spoken to before, and they asked me if I would like to meet a man I had never met. I was told this could be of real interest to me, and on hearing his name I thought the same. Julian Assange – the Australian provocateur, a Scarlet Pimpernel for our digital age – wanted to talk.

Julian was at London's choice venue for hard-bitten hacks and war correspondents, the Frontline Club in Paddington. Heading there, I found him holed up in one of their rooms, being interviewed by CNN. Nervous and a little self-conscious, he was unused to the media spotlight and here he was being asked about a set of documents his whistleblowing organisation, Wikileaks, had just released: a cache of military reports that exposed the truth about America's war in Afghanistan.

Julian had some of the most controversial secret documents ever to find their way to the light of day. Millions of files from the US diplomatic and military operations overseas that had been leaked by US soldier Bradley Manning. And I was there, as the editor of the London-based Bureau of Investigative Journalism, to see if my outfit could have a peek. Julian, interested in the Bureau's ability to make documentary films, was keen to see if the contents of another set of files, this time the Iraq War military reports, could end up on TV channels the world over.

It was a treasure trove of documents that proved there were war crimes and human rights abuses, incompetence and intrigues on the part of the US military in Iraq. And Julian gave it over, countless classified military files, on a USB stick in a Lebanese restaurant near Paddington station.

As I got to know Julian, his appearance changed considerably as the world focused more and more on what he had leaked. He lost weight, dyed his hair bleach blond, took on a lined tiredness. But there was one thing that was constant: his bulletproof vest. He feared a CIA attack, that he was an assassin's target, and I couldn't help but look at the bulky blue vest and think: headshot.

A few days after giving the Bureau the files on the Iraq War, I was told by Julian to download a messenger system called Jabber. It was an encrypted service that lets people talk relatively securely. So that night, I logged on and began a conversation with one of the most controversial men in the world at that time.

Caught in the green-blue glow of a screen, I was told to download certain applications, and his terse words guided me through a portal I had never known existed. I felt ashamed at my technological illiteracy. He told me about TOR, a system that allows its users to search the internet without their computer's address being revealed. One that lets you look at websites untraced, because TOR wraps your servers' information around other servers' information, hiding you behind peeled layers of anonymity, like an onion.

Clearly, the head of Wikileaks needed the anonymity that TOR offers, just like investigative journalists do. But some others do not. Others use TOR not out of need, but desire. For many things lurk deep in the hearts of men, and if you give them a tool to hide their identities they will use it.

Within minutes I had access to sites that sold things like $20 syringes full of HIV positive blood, a vendetta's stabbing tool. Where you could buy a soldier's skull from Verdun for $5,000. Where fake euros cost a fraction of the real price. There was even an encyclopedic portal of links to ugly places I did not care for. Websites like 'Pedofilie' or 'Boyloverforum'. Places that offered you the chance to commission drug experiments on homeless people. There was the opportunity to purchase snuff films made to order. You could buy the contact details of 'crooked port and customs officials' or 'discreet lawyers and doctors'. And then there were the sites that promised assassinations.

'Unfriendly solution' was one. The text was chilling, if a word of it was true.

'I will "neutralize" the ex you hate, your bully, a policeman that you have been in trouble with, a lawyer, a small politician . . . I do not care what the cause is. I will solve the problem for you. Internationally, cheap and 100 per cent anonymously.'

The text continued: 'The desired victim will pass away. No one will ever know why or who did this. On top of that I always give my best to make it look like an accident or suicide . . . I don't have any empathy for humans anymore. This makes me the perfect professional for taking care of your problems . . . I'll do ANYTHING to the desired victim.' The price varied. For non-authority, 'normal' people the cost was between $7,700 and $16,400.

Another site, 'The Hitman Network', offered a team of three contract killers in the US and the European Union. They charged $10,000 for a US hit, $12,000 for an EU one. They had two rules: 'No children under 16 and no top 10 politicians.'

Of course on the dark web you have no idea who is and who is not genuine. It could just be some fat guy called Bob sitting in his underpants in his mother's loft in Illinois. Or a cop. After all, the US's Bureau of Alcohol, Tobacco, Firearms and Explosives has acknowledged it runs agents who pose as fake hit men, men who wear the jewellery, sleeveless tank tops and facial hair of biker gangs, entrapping those who seek guns for hire.[35]

In a twist of irony, it was even alleged that Ross Ulbricht, the supposed founder of Silk Road – a TOR portal that sells all manner of narcotics, drugs and illegal services – commissioned the murder of six people through hitmen he had contacted on the internet.[36] Nobody was actually murdered, although the FBI did say they had faked the death of one former employee of Silk Road and claimed they had convinced Ulbricht the murder had taken place. Ulbricht reportedly wired $80,000 to pay for the hit,[37] and was sentenced to life in prison.

But some of those on the dark web might, just might, be genuine. Because there are definitely men out there who earn a bloody living using the gun in an ugly and vicious way, and there have been for some time.

The first gun assassin to enter the history books appears to have been a Scotsman – James Hamilton. Hamilton, a strong supporter of Mary, Queen of Scots, set out to kill James Stewart, 1st Earl of Moray and regent of Scotland for the English Crown. The sniping Scot, in support of Mary, had fired the lethal shot from an archbishop's window, through a line of washing. Since then, we have seen world leaders gunned down in their cars, on motel balconies and in theatres, a single bullet spinning a nation into a state of mourning, spurring on a legion of conspiracy theories, and even triggering world events that have claimed the lives of millions.

So President Abraham Lincoln was shot in the back of the head at Ford's Theatre in Washington, DC by actor John Wilkes Booth, wielding a Philadelphia Derringer with a black walnut stock inlaid with silver.[38] Archduke Franz Ferdinand was gunned down in the streets of Sarajevo in 1914, wearing such a tight uniform that some speculate it even helped speed his death.[39] And a host of others have fallen to the assassin's bullet, Tsar Nicholas II, Mahatma Gandhi, Yitzhak Rabin, President Kennedy, Dr Martin Luther King Jr and Malcolm X among them, victims of the powerful political symbolism the assassin's bullet delivers; potent propaganda in a bloody deed.

Like the best propaganda, the world of the assassin does not easily show its true face. But glimpses of it fascinate and endlessly inform the subject of films and dramas, which, in a way, seems ironic when you read that the father of Woody Harrelson, the star of *Natural Born Killers*, was actually a contract killer. Charles Harrelson was given two life sentences in 1981 for the assassination of district judge John H. Wood, the first murder of an American judge in the twentieth century.[40] The killing was carried out with a high-powered rifle in return for $250,000.[41]

Other shadows emerge from this world, such as the thirteen-year-old hitman working for the Mexican drug cartel. In 2013 Jose Armando Moreno Leos confessed he had participated in at least ten homicides, hired for his skill at shooting a high-calibre weapon. He was freed, because the Mexican constitution prohibits the incarceration of those under fourteen, but a few months later Jose too was found murdered, execution style.[42]

Or there is the story of infamous Russian hitman Alexander Solonik, better known as Alexander the Great, who confessed to assassinating a string of Moscow underworld figures in the 1990s. He had a unique skill: he could shoot with both hands.[43] Or the time when, in 2011, Indian police arrested one of their most notorious contract killers, Jaggu Pehelwan, a man believed to be behind the deaths of over 150 people. He charged, it was claimed, between $19,200 and $50,000 for each kill and had even done a deal agreeing to two dozen murders for $307,000.[44] He died as he lived, shot by rival gang members in 2012.[45]

Even Britain, with its hard gun laws, has had its share of hitmen. Santre Sanchez Gayle was Britain's youngest, just fifteen when he shot a young mother for $300 in Hackney in London.[46] He was ripped off. Researchers found the average cost of a hit in Britain between 1974 and 2013 was just over $23,000. The highest was $150,000, $300 easily the lowest.[47]

The assassin's bullet, of course, is used for many reasons. For the assassin, cash has to be their sole motivation. 'Don't take it personally,' is their cinematic shrug as the gunman leads a cowering accountant into a secluded wood with a shovel.

But what happens when the killer's art is used for a darker purpose – for the pursuit of power?

It was a question that led my focus and research away from the depraved minds of psychotic killers and guns-for-hire towards the more calculated horrors of gangland murderers – the domain of the criminal with a gun.

6. THE CRIMINALS

A surprising video and gangland killings – El Salvador – meeting 'the Shooter' and drinking to the Beast – tea with a spook – secret graves and grave secrets – Holland and Ecuador – personal trauma of guns recalled – memories of Papua New Guinea – a gunpoint mugging and hard justice seen

The video was of a New York that no longer exists. Mean streets against a dripping industrial grime. So far removed from today's Manhattan's chic and West Village bearded hipster irony, it wasn't even New York. The producer called the song 'Gotham Fucking City'. The three-minute, twenty-second video was the story of Paris Lane's death – a suicide captured in song. A eulogy in gritty rap by the singers Smoke DZA and Joey Bada$$, with lyrics as hard as the story it was telling. The video told me a different story. It did not tell me the one I knew, how Paris Lane had killed himself in that New York lift lobby because of depression or demons or other silent assassins. Instead, it showed that he killed himself because of a drugs deal gone wrong. Having read the comments under Paris's suicide video, I had searched some more and found this rap video. It was not a story I had visualised. Only the ending was familiar. The last frames in the music video showed the image of a young man standing before a closing lift, a gun in his mouth, but this scene was intercut with others: young thugs with guns seeking vengeance, lurking at the tower block's doorway.

The hooded men were shown in the film being robbed twice. The first time Paris had seized their drugs. The second time he had denied them their chance at dispensing a brutal form of street justice. Perhaps Paris knew, schooled as he was on the city's sidewalks, that he couldn't get away with what he had done, so he chose a different road.

The comments below the online video were illuminating. 'R.I.P. paradise he killed his self before everybody he robbed killed him first,' someone had written. Underneath it another said: 'Supposedly he wanted out of a gang and they were going to kill him so the last thing he did was say goodbye to his girl before leaving on his own terms.'

Perhaps it wasn't a suicide at all. Not in the strictest sense, at least. Rather just the bloody ending of a gangland disagreement. Something that had gone so far south that the only way for Paris to find a way out was to put gunmetal in his mouth.

Truth is sometimes hard to see. It's even harder when there's a gun in the way.

I began to search for the people who wrote the comments. They had indurate street names. One emailed a reply but I won't say who because in the world of street gangs, that can be as good as chiselling a name on a gravestone. He told me Paris was in no mood for suicide.

'i [sic] can't tell you who he was in trouble with. Some guys from the hood. I heard his old gang friends tried to get him but nobody talks about this shit.'

I looked at five of those tight words and realised I had been expecting them all along. *Some guys from the hood* – because you can't travel into the closed world of the gun without meeting the gangs who run the city's night streets.

There are an estimated ten million criminal gang members in the world, and it was clear I had to meet some of them.[1] To really understand the world of the gun I had to cross over into the criminal badlands.[2] The question was: where?

Considering this, my eye had fallen back on Latin America, because one in seven of all homicide victims globally happens to be a young man aged between fifteen and twenty-nine and living in the Americas.[3] And in Latin America 30 per cent of all homicides are linked to organised gangs.[4]

At first, Mexico seemed a good choice. Mexican cartels' firearms of choice include .50-calibre sniper rifles and a Belgian bulletproof vest-piercing pistol they call the 'cop-killer'.[5] But I had once reported on just how badly the gangs there tortured people, and I feared them. Nothing is worth being skinned alive for.

Then I read how El Salvador had 60,000 gang members in a country with a police force of just 25,000. About 80 per cent of all murders are with firearms; there were over 3,000 gang-related deaths in 2009 alone.[6] This worked out at a thousand more gang-related deaths in one year in El Salvador than in the whole of the US, a country fifty times its size. My curiosity was piqued.

I learned that El Salvador has two main gangs. There's the 18th Street Gang, or the 18. Having originated in California, the 18 had grown to become an international gang with 65,000 members in 120 cities. They are renowned for rituals like having large tattoos of the number 18 inked across their faces, or initiating gang members with a vicious '18 seconds' beating'. Their rivals are the Mara Salvatruchas, the Maras, or the MS-13. They have over 70,000 members worldwide and are heavily involved in black-market gun sales, human trafficking and homicides, especially of law enforcement officers.

On 8 March 2012, the leaders of these two gangs in El Salvador had called a truce to the bitter war they had been fighting. In exchange for peace on the streets, they wanted a promise from the state for an improvement in the living conditions of their gang members in jail. The government, in an unprecedented act, was said to have negotiated with the gangs and come to a deal.[7] Whatever the reason, a truce was struck. Almost immediately, homicide rates dropped off a cliff; the usual seventeen killings a day fell to an average of just over five a day.[8]

It was a laying-down of arms that created an opportunity I felt I could seize. Often gangs can be so caught up in a vortex of their own violence that even talking to them drags you into their chaos, but in El Salvador things seemed to have calmed down enough to get a glimpse of gang life close up. So it was that I decided to travel there to see what gangs and their guns can really do to a country.

The country was preparing for elections, and the humid streets had taken on a festive air. Banners fell from windows, and vibrant flags in red stars and blue stripes fluttered in the sporadic breeze. The occasional political advertising car drove past, loudhailers promising a better future – at least a better one than this. All the parties were running on the same ticket: security. But political promises had been broken so often in this Central American country it was hard not to shake your head at the cars passing.

I was not here to report on politics, though, at least not directly. I wanted to meet gang members and had asked for the worst my American-Salvadoran fixer could manage. He said he would not disappoint.

He arranged for two to come to my hotel in a crumbling northern district of the languid capital of San Salvador because the alternative was for me to travel to them. He knew I wanted to talk to them about guns and death, and as words quickly transform into actions in these men's lives, so we met where an illusion of safety could be kept.

They arrived on time. They were both Maras, two of an estimated 25,000 in Salvador. They were in their late twenties, old for some of the gang members here – many died much younger. The taller of the two was called 'the Shooter' – he was dressed in a neat black polo shirt, his slim frame and supple hands belying the fact he had killed so many. His friend was quieter, softly spoken; his blue striped top and neatly cut hair made him look like a shopkeeper, not a killer.

We shook hands and sat beside a swimming pool lined with rich lotus flowers, and thumb-sized insects buzzed among their sweet petals. They ordered strips of beef and corn-brown tacos and drank from sweating glasses of Coca-Cola.

'When it comes to having to bury the body,' the Shooter said, pulling apart his meat with a fork, 'you don't want a long tomb, so you cut off the head and the arms to bury the corpse. We bury it like an animal, like it's rubbish. The only reason we bury the corpse is because of the police and the evidence. Otherwise, we'd just leave it to the dogs.'

There was no small talk. Within minutes they were explaining how the gangs here sold crack cocaine, prostituted twelve-year-olds, extorted businesses. They ran brothels where you paid a set fee before you went up with the girl of your choice. When I asked how many of them had spent time in prison, they just laughed.

They carried on eating and, in between mouthfuls, explained how they treated their enemies without a trace of humanity. They once killed a three-year-old because his father was not at home for the punishment they wanted to dispense. They would kill you for any reason: they'd shoot you if you were gay; they'd kill you just because someone asked them to, without question. And now they insisted that those who wanted to join their gang killed too.

'Before, when you wanted to enter the MS, you had to be kicked, beaten for thirteen seconds,' the Shooter said. 'But now we give you a name, a target and you have to shoot them. We do this to show you have the balls to confront your enemy and to enforce your loyalty to the gang.'

An insect landed on his face, but he did not swipe it away. 'We give these targets to boys, twelve, thirteen years old. The victim could be anybody, but it is normally an enemy: a guy from the 18 gang. It could be someone we have kidnapped, someone who has stolen money from the gang. A Mara even: a snitch.'

He carried on talking, pausing to scoop rice and beans into his mouth. 'Last year we had a guy who wanted to be in the gang, and his target was chosen at a nearby school. Next to the school was a police station. In this case the new guy waited for the 18 gang member until his class was over and then, as the target left the class – "boom" – he was hit. We chose it there because the recruit was forced to do the hit in front of the police.'

The killer was fourteen years old.

Like other gang members I had met, the Shooter enjoyed telling me these things, because he took pride in how they killed and he mistook my interest for admiration. I asked what happened if someone didn't follow through with a killing.

'If you don't do the hit, then they'll kill you, as you violated the code.' I imagined a thirteen-year-old losing their nerve at the last moment, refusing to pull the trigger. And that being the last good thing they ever did.

'If someone was in the MS and then told the police about a killing, what would you do?' I asked. I knew the answer, but sometimes you ask obvious questions to hear what people have to say.

'It would be a tougher, more brutal death,' the Shooter said. 'We'd take the snitch out to a secluded place and tell them that they were going to kill an 18 member. Then we'd hack him to pieces with a machete, not using a gun, because a knife is a harder way to go.'

His voice was low and full of menace. 'I'd kill a whole family, you know. I'd kill them because if I just kill one and the others grow up, they'd kill you.'

In the still of that space the words sunk in. The air had grown sauna-hot, and the insects had started biting. I was struck by how young the gang members were. Perhaps the gun makes their killing easier – a knife might be too personal, more difficult to do. But a gun? You turn your face, pull the trigger and it's all over. Guns just made life cheaper.

'Do you have nightmares?' I asked.

'Only at first, after the first kill, I had a few days like that. But now? No. It has just become usual. You just get used to it. You become anxious if you aren't killing.'

Then, like a shopping list, he told me what guns they used. 'We have the usual: AK47, M15, .45 – you know. Every type of calibre. We need to be well armed.'

So the police armed themselves against the gangs, and the gangs armed themselves against the police. With such logic, it was no wonder that there were over 225,000 illegal guns in this country.[9] And the vast majority were simple, inexpensive weapons – cheap revolvers and semi-automatic pistols made up 78 per cent of the guns seized by the government in 2011.[10]

'A few months ago I got a call. Someone had got their hands on an AK47 with a telescopic sight. It cost about $3,000,' he said. Of all of these weapons, the most coveted was the AR15 semi-automatic. They brought these in from Texas, and they could fetch as much as $5,000 each.

The Shooter's conversation about guns was measured and cold. Their cost, the way he got hold of them, the things he used them for: he spoke without passion. The gun to him was just a tool. In fact, the only time either showed a spark of life was when they described how, after each killing, they would drink to the Devil.

They splayed out their hands in the signal of the MS, little fingers and thumbs spread out like horns.

'To the Beast,' they said, and they finished their drinks.

The middle-aged man did not look like someone who worked for the intelligence services. His swept-back, receding hair, neatly trimmed moustache, blue jeans and cream short-sleeved shirt gave him the appearance of a used-car salesman. But he was one of the best-informed agents on gangs in Salvador, so I accepted his offer of gritty coffee under the photographs of his fallen wartime comrades that hung upon the grey walls of his dimly lit office. I pulled up close to catch his quiet words.

'Basically, this country has been overtaken by organised crime,' he said, our cups delivered by a secretary wearing a tight skirt that caught his gaze, 'and weapons play an important part in this because they give the gangs power. They get their guns from Honduras and Guatemala – illegally trafficked in. They come in via corrupt border officials.'

He had been in the Special Forces during the war, on the side of the right-wing government. His war was not really over. In his mind, the forces of evil, be they the anarchists of the left or the cartel-backed gangs, threatened his way of life.

He drank the hot coffee fast and put down the cup neatly. The main problem, he said, was the guns left over from the civil war. Here, in the 1980s and '90s the US had flooded the country with countless arms, supplying the El Salvadoran military alone with over 30,000 M16 assault rifles.[11] On top of this, there were an estimated 100,000 pistols and rifles kept behind by the left-wing revolutionaries.

You believed him, not because he was likeable but because he was precise. His napkin was neatly folded beside his aligned cup of coffee. His pen and his phone laid out in parallel next to each other. The private habits of a man say a lot about him.

He had asked me not to use his name, and then, as we talked, he grew conspiratorial, and his fingers fluttered to his moustache. Some of the gang's guns also come from the military, he said. 'About 60,000 rifles were handed into the military armouries after the war, but even they are not secure. And some soldiers are corrupt, too.'

He began to rattle his pen angrily against the handle of his mug. El Salvador had, he felt, become a country where it was questionable if the state was truly the main power centre. It had become so linked to a web of illegal activities that at times the government itself was the criminal enterprise. 'The gangs are a violent force to the state, and they should be treated like a national security issue. They respect nothing, they have no codes, discipline. National laws here are not as important in some areas as gang laws,' he said. 'The unspoken laws that hold gangs together are harder, more enforced, than the national laws.'

He had a point. Certainly, talking to experts and gang members around the city, I had found that the truce between the gangs was not working – that the number being killed was more than publicly stated. People said the leaders had used the truce to put themselves on an equal footing with the political leadership. The truce had given the gangs legitimacy, even.

'The reasons for joining a gang today, after the truce, are changing,' the agent said. 'Gangs have become more politicised, more strident, and this changes the type of people in the gangs. The gangs themselves are transforming – the government struggles to contain their communication networks, their structures, their armaments. Here the gangs have evolved into something else. Perhaps even the fifth estate – the criminal estate.'

Jailed gang leaders had been granted generous concessions by the state. Videos had surfaced of trainers visiting El Salvador's prisons, teaching incarcerated gang leaders combat tactics. Gangs were even now renting policemen's uniforms and badges to set up roadblocks to rob passing vehicles, and worse.

What upset this man so profoundly was that he knew, deeply, that the line between the state and the criminal underworld had become blurred, that all around him people were just looking the other way. And he was right.

A few days earlier I had been shown a scrub patch of dirt and dust in the centre of San Salvador. It was less than half a mile from the capital's main police station. Gang signs were spray-painted upon the surrounding walls, and a broken car was parked beside it, filled with young soul-stained boys: the ears and eyes of the gangs on these shantytown roads.

My guide, a man who had seen violence up close, nodded to the illicit grave and said he knew of others like it. Children, street vendors, the elderly, rival gang members – these were just some of those found in graves like this one. As the days passed, people were to comment that the mutilations that appeared on these bodies had not been seen since the 1980s, when El Salvador's vultures had grown fat upon the unholy carrion of war.

These illicit graves were disturbing for another reason. The only reason for a truce between the gangs and the state was to reduce the number of homicides. Now it was apparent that, rather than dumping the bodies on the streets, the gangs were just burying them in secret.

But you do not have to go to illicit cemeteries in Salvador to see the impact gun violence has had. Up from the graveyard we passed other scenes of desolation. Litter-swept homes gutted by fire, crumbling masonry on streets lined with fly-blown mounds of stinking refuse. The sharp smell of oil and decay lingered. And on each side graffiti scrawls on faded walls marked the territories of those who controlled these streets.

It seemed everything in Salvador was laid with rolls of sharpened razor wire; a country under siege from the inside. People lived in homes high-gated and high-walled. Listless security guards cradled shotguns at even the smallest convenience stores. Old women sat behind iron grills and stared into the streets with fearful eyes. Everywhere was *peligroso*, dangerous.

Later that day I travelled out to a district in the city run by the 18. So absent was the state in the streets of this small barrio that the gang was the only law here. Without being invited in, I would have been killed.

I had been told that the young men, in their basketball tops and wide-rim caps, liked to eat fried food. So I handed over $50, and someone got us oil-blotted cartons of greasy chicken wings and thighs, and we sat at a makeshift table on a dirt track. Beyond, a dull red road led up to a bank of lush foliage, the silhouettes of men walking listlessly on its rutted surface. Ants crawled everywhere.

I edged around the thing I most wanted to know: how their gang ruled the lives of others here. I had been told not to ask questions that might anger them, so I began to talk about my own experiences of violence, and the young men listened. Then they began to talk, because, compared to them, I had seen nothing, and they needed me to know that.

'I don't have to go to Iraq or Somalia; that violence I can find here in my home,' said Mario, the captain of this gang. He had a rosary around his neck, and on the cross hung a small silver figure of the baby Jesus with outstretched arms. Mario's own arm was marked by a spiderweb of gnarled white flesh from his wrist upwards – the scarring of a bullet. He began to speak about power.

'The 18 have more control, more respect than anything else here,' he said. 'We have the control of the land here, but the fact that everyone has a gun means we will always be protecting ourselves from others coming in.'

He talked about the collusion of the state in the gang violence, and you realised just how far things had slipped. 'If the army captured you, they would take you to MS territory and leave you in the middle of it, with a gang. They'd take your cuffs off you and leave you to them.' Some policemen would even rent out their guns for the weekend if you wanted to do a particular hit.

I asked him what he would do if the government tried to curb their powers.

'If they attack us, we will respond,' he said and left the unsaid hanging in the air. This was it: what happens when a gang's grip on violence supersedes a corrupt state's ability to curb that violence. Broken lives, ruined infrastructure, abandoned hope.

There were parts of Salvador still functioning, but when a country starts negotiating truces with criminals with guns, you wonder where it will lead. Then when the government starts ignoring the clandestine cemeteries that fill up during that truce, you know that the path to righteousness has long been lost.

I've been held up at gunpoint three times in my life. I've also been shot at twice. The reasons I was shot at were indiscriminate, so I didn't take those times personally; I just happened to be in the wrong place at the wrong time. Being held up at gunpoint, though, was intimate. Those moments of rushed demands and ugly threats have stayed with me longer and touched me more deeply than most of the horrors I've seen. Something in me broke a little.

The first time I was held up I was twenty years old and in Amsterdam. I was with friends, and we had staggered from a bar, a wraith of sweet-smelling smoke curling from the door behind us. We had walked into a Dutch world of shining cobbles and wet canal railings, cheap neon lights and hooded shadows. That is when the two men approached. One was white, a South African, his face taut against his skin, a shade of mottled green in the light. The other was a black guy, silent, his background unknown. He held a tightly wound package in his right hand.

'See this,' the cadaverous white man said to me, his mouth thin, spittle on his lips. Then he demanded money, he said they had a gun. I asked him if I could see it, as I did not believe their threats, and they unwrapped the package until I could see gunmetal in the dull light. So I told them that I didn't have any money for them, and the South African let loose a thin line of spit on the ground and told his friend I wasn't worth it. They walked away, their jackets furling in the wind.

The next time I was held up was a few years later. It was too similar to the first time to warrant the retelling. A late-night street; a man with a gun; a realisation that I didn't have any money. This time it was in Guayaquil, a languid heat-stamped town in Ecuador.

So I took to carrying a $100 bill with me because I felt it was better to have something rather than nothing in my pockets to give these midnight vultures.

The third time? Well, this one stuck with me.

It was 1996 and, at twenty-three years old, I was brimming with confidence – perhaps too much so. We had not been out of the Papua New Guinea village for more than an hour when we were robbed.

The slope leading up to the thicket of trees that lined the upper levels of the rainforest mountain were bare, save for the willowing grasses and the low clumps of scrub. Beyond was an endless spread of trees, impenetrable and secret – a green hell for the stranger to these lands.

On either side of us ran a troupe of excited Papuan children. They had followed us from the village below and were singing back the lyrics to *The Sound of Music* I was trying to teach them.

I was with a friend, Robin Barnwell, a talented BBC filmmaker and someone I had known since childhood. We had decided, one drunken night in London, to pool our last savings, borrow some more money and have an adventure. We wanted to go night hunting, dive on wartime wrecks, hike at the edge of the known world. The best way to have that sort of trip was to go to Papua New Guinea. Outsiders had only reached its Highlands some fifty years before; ancient tribes were still living in the depths of these green mountains, and people existed as their ancestors had done. Customs and traditions here were only lightly touched by the modern world; cannibalism, tribal warfare, animism – all lay just beneath the surface.

We were young and, walking so blithely up that mountain path, a little foolish.

'Doe, a deer, a female deer. Ray, a drop of golden . . .' the children's voices lifted and were filled with laughter. Flashes of teeth and bubbling excitement; they tripped and gamboled. Then, suddenly, they fell silent and stood.

Ahead was a small group of men. They were, like many here in Papua New Guinea, wearing tight and elaborate wigs. I waved. It seemed to be a small hunting party coming home, but they had nothing to show for it. No short-beaked echidna or a long-fingered triok or whatever else they caught up there. Then I noticed something wasn't quite right, and they began to run hard at us, their muscles taut under their dark skin. The children – the *pikininis* as they call them in Papua New Guinean Pidgin English – squealed and scattered.

Then, in a stumbling, scrambling grasp of what was happening, we realised these men were part of a 'Raskol' criminal gang, the name for the wild groups in the Highlands of PNG. They were carrying two long rifles and a sharp-edged machete, and the three of them came up close, levelling their guns.

'We will kill you!' they screamed.

The words tumbled from them. They were demanding money. We knew that Raskol robberies normally ended in violence: they won't just rob you, they'll end you. So we bowed our heads and fell to our knees, and they grabbed our rucksacks and money pouches. Passports, clothes, everything: all gone. Then, as quickly as they had arrived, they left. The children beside us began to cry, but we were thankful to be alive.

Then we heard screaming, and things took a turn for the worse. They were running back – one of them had levelled his rifle again, and the space was filled once more with snarling and screaming. They ran straight at Robin. He swivelled and, pushing away, bolted back down the path. I remained. I didn't want to die with a machete in the back of my head. Instead, I began to unbutton my shirt.

I recalled reading somewhere that it's harder to kill a naked man than a fully clothed one. That human flesh somehow humanises people. So I took off my shirt, button by fumbled button, without really taking it on board that the man who stood before me pointing a loaded gun at my head was totally naked, save for a quivering penis gourd. My nakedness was, clearly, not going to sway any intention of violence.

Robin was dragged back up the track, a knife close against his throat. And then, in all the screaming and threats of violence, they handed us our passports back and, with an angry push that sent us tumbling into the dirt, they disappeared back into the green screen of forest. I guess they thought taking our passports might cause our governments to come here with even more guns and attack them.

With just our trousers, shirts and now passports, we looked at each other on the side of this Papuan mountain, miles from anywhere, and considered our options. This was not a good moment, but we knew it was best to leave before they came a third time. So we walked back to the last village.

There, the village headman of the Hulis, a stocky man in a tweed jacket and grass skirt, met us; he was pushing a long reed through a hole under his nose. Some of the pikininis from his six wives had already told him what had happened, and a messenger had been dispatched to a local police officer in the valley below.

'It was the Engans. They are always doing this sort of thing,' he said to us in Pidgin. 'They are just a bunch of idiots.' Or words to that effect, because Pidgin English insults are hard to translate.

We were not surprised. The Enga tribe had a long history of violence. Walking had shown us the sort of low-level war they were waging. Houses had been burned, trees reduced to ugly, jagged stumps. It was a type of conflict fought in weekly battles on football-pitch patches earmarked for combat. Here, like medieval warriors, each side would stand and hurl weapons at the other. This violence was devastating. One tribe, the Mae, had an estimated annual death rate of around 300 killings per 100,000, almost a hundred times that of the US.[12] It was once estimated that two-thirds of the men of another tribe, the Gebusi, had murdered someone.[13]

Then there were the violent raids upon neighbouring territories that took place every few years; some tribes even raided a dozen times a year. What they did during their slaughter, when the red mist descended, was graphic. Some tribesmen believed they could boost their masculinity by eating the brains or even the penises of their enemies to absorb their strength.[14]

Yet a fundamental shift had happened. Once these killers used to have bows and arrows, but today they have shotguns and high-powered rifles. The introduction of guns into this tribal culture changed everything. Before, the traditional weapons of clubs and arrows killed few, but their new weapons massively increased the numbers killed. One report even said the spread of semi-automatics was 'out of control'. And the Southern Highlands, where we were now, was considered the worst place of all.[15]

I guess we were lucky not to have been executed. But now, in the sanctuary of the village, there was not much else for us to do but wait. So, a tattered and much-used leather ball was brought out, and we played a terrible game of basketball and, sipping a warm bottle of Coca-Cola that the local village store sold, we sat under a reddening sun.

The roar of an off-road vehicle disturbed the quiet. The track leading up to the circle of mud and thatch huts was deeply scarred by the mountain rains – ravines in the road. But the vehicles here were as rugged as the barefoot tribes, and the police car pounded up it with ease, braking in a squeal of metal in the middle of our game.

A huge man, his shirt straining at his barrel chest, got out of the car. A Papuan policeman, he was the largest person I had seen up there. He had a thick moustache and wore mirrored aviator glasses, like a 1980s New York cop. He was chewing betel nut, the seed of the areca palm, and his mouth was a red smear. It gave him, along with his heavy black combat boots and the oiled pistol on his hip, a dangerous air.

He listened to what had happened and climbed onto the bonnet of his vehicle. His voice was forceful and deep. 'If these white men don't get their bags back in the next twelve hours,' he said in the local dialect, 'I will burn down all the villages in this valley. Each and every building.'

And that was it: Papuan law enforcement. We were horrified: our uncalled-for presence here had resulted in the threat of a mass burning of villages. We tried to protest, but he ignored our pleas. This was the Highlands, and this was how they did things: a form of retributive justice.

His threat worked. Twelve hours passed, and our bags turned up, slipped back to us under the cover of night. A neat cut of our money had been sliced off the top, but we were told not to bother about that. The case was settled.

Looking back on it, perhaps I now see how so much about guns and crime and policing lay in that small episode. The terror of the robbery and the small humiliations that come when you are faced with intimate lethal force. The breakdown of the rule of law in remote and impoverished armed communities. The state's exercise of power through a stronger, better-armed force and the casual dispensing of justice.

It was an incident that helped frame my thinking, too, when I was to shift my research away from the illegal use of guns by killers and criminals to looking at those who use their guns in the name of the state: the police.

7. THE POLICE

The trouble with police embeds — the gun-sniffing dogs of South Africa — chasing gangsters in Cape Town's slums — talking to an American police sharpshooter — understanding their warrior cops — the Philippines recalled — when the police murder — the death of an activist in Mindanao and beyond

One of the challenges as a journalist, as a writer, is meeting the police.

Generally, when you embed with the army in a conflict zone you are sent in to see some 'action'. Politicians and military press officers want to show the world that their troops are at the hot end of things, guns at the ready, doing the job they were sent in to do. Journalists have made careers from these high-octane embeds, risking their lives in bloody campaigns, often acting as the liberal conscience of a nation at war.

Hitching a ride with the police, though, is different. Rarely do you get on a hard-ended raid. I have been on embeds with cops in some of the most criminally active parts of Latin America: El Salvador, Honduras, Colombia, Brazil. And each experience, oddly, felt the same.

It begins with a tense shaking of hands, because policemen distrust anyone who is not one of them. There is the required donning of a bulletproof vest; the faint promise that something might happen. Then, after a briefing, there is a drive and the inevitable crackled call to a crime scene, sirens flashing. The destination, though, is reached as another police unit has apprehended the criminal, or a body is already lying in the street, or it was a false alarm.

The basic truth is that policing is deeply political, and no police media officer wants a journalist seeing anyone getting shot by their boys in blue. So they send you to places that once were terrible but have since been tamed. They get you on community policing initiatives. It's all heart and minds, not blood and gore. And it's not just the censorship of a press officer that's at work here. The reality is this: most police don't really use guns that much.

With the people I had looked at so far – mass killers and criminals – guns were central to their actions. But I have found, over time, that police forces often have a much more complex relationship with guns. Some forces are very weapons-focused; some concentrate on intelligence-led policing; others use different forms of restraint, like Tasers. A policeman's gun, when used appropriately, is not used in attack, but rather in defence – upholding the law, not imposing something.

Of course, police forces carry guns, and lots of them. Of the nearly 1 billion firearms worldwide, law enforcement organisations have about 25 million – about 1.3 firearms per officer. And the bigger the force, the more guns there are. China's police have an estimated 1.95 million; though its sheer size means – at about seven firearms for every ten officers – they have fewer per officer than the global average. India's police have fewer guns in total – 1.9 million – but more per officer – around three for every two policemen.[1] But neither country boasts the most guns per officer: that prize goes to Serbia with two per officer.

Such figures, of course, only take averages into account. In the US there are about one and a half firearms for every law enforcement officer. But the average officer in the Federal Fish and Wildlife Agency has almost six weapons, such as a Sig Sauer model P220 semi-automatic pistol or a Remington 12-gauge shotgun.[2]

What interested me were those police worlds where the gun was very present and where that presence transformed policing in a profound way. So I looked at three areas where firearms were at the forefront of law enforcement: police gun units; their use in para-military-style raids; and the use of the gun in the police abuse of power.

The dog was visibly excited. It knew something was up. His handler threw sand in the air to see the way the wind was blowing and then unclipped the collar. The brown-flecked Border Collie was off the leash and it scurried from tyre to tyre, laid in a neat row behind the school. At the third one it swivelled on its hind legs and sat down. The handler walked over and, patting his dog, reached into the darkness of the rim; with forefinger and thumb he lifted out a Glock pistol. The dog started to bark.

South African sniffer dogs are trained for two purposes: to find drugs or explosives. The former are experts at sniffing out cocaine and marijuana. The latter are like the dog that was performing before me – able to smell the remnants of an explosive blast, detecting cordite and primer. Gun dogs in a unique sense.

In South Africa such dogs are in constant use. In the south-west the Cape gangs have an estimated 51,000 guns. And this is just the southern tip of the country. Nationally there are as many as 4 million illegal guns; in some areas I was told, with the right contacts, a gun is easier to get hold of than a glass of clean water.[3]

The effects of this profusion of illegal weapons is clear: at one stage 15,000 South Africans were dying from gunshots a year.[4] Police and their gun dogs here, frankly, have their work cut out.

This is why, having met the doctors whose jobs it was to patch up the wounded, I had arranged to meet a police squad dedicated, literally, to sniffing out guns in one of the most violent parts of the world. The dogs here, though, were not being used to search the gang borderlands of Cape Town. Rather, we were at a school. Drugs had been reported being used openly in Hoërskool Bonteheuwel, a small college fixed in one of the poorest neighbourhoods of the Cape Flats. A gun had also been seen, so the armed police had arrived during a morning congregation, announced a raid and told the pupils they were to be searched.

The gun sighting was no surprise. Children as young as fourteen were being arrested in the Cape Flats on gang-related murder charges. Violence had spiked; in 2013, 12 per cent of the 2,580 murders in the province were gang-related, a sharp rise on the previous year.[5] The cops were taking no chances. Each child was ordered to put their hands on their heads and, row after row, these young coloured kids were frisked.[6] The children rose silently, one by one, to have a policeman's hands search their skin-and-bone bodies. Large white men went over trouser turn-ups and polo-shirt collars, hunting for drug wraps, flick knives, guns. Some of these kids had said goodbye to childhood a long time ago.

The police worked quickly under a faded wall painted with the words: 'If you do good, no one remembers. But if you do bad, everyone remembers.' But school proverbs only do so much; it was the influence of fathers and uncles that the police really had to deal with. In the 1980s, during apartheid, there had been a major relocation of coloured and black people from Cape Town's inner city out here to the Cape Flats and its surrounding townships. The move had spawned a violence that now spanned the decades and ultimately had given birth to a breed of gangland criminal here that had nothing to lose.

In the quiet of that morning, I could see through the windows an obese woman waddling past, her bulk generously covered in striped fabric. She was pushing a broken pram and looked down at the littered floor, not caring what was unfolding across the road. This was not the first time armed police had been at this troubled school.

Inside, an expressionless officer pulled aside a boy. The youth had tucked his trousers into his socks: a gang sign. The inspector held the child's shoulder as the dogs scurried around his feet, noses alert. They sniffed at the boy's legs and turned back to the chairs. No drugs or cordite here, and apartheid's ghosts stood watching from the shadows.

I walked outside and went over to a gaggle of five Cape Town Metropolitan police. They were idling, their Glock pistols untouched in blue plastic holsters. Each had been in the force for twelve, thirteen years, but, despite the violence of the Flats, none of them had ever fired these pistols at anyone. It was turning into one of those police embeds.

Perhaps sensing the disappointment in my eyes, an inspector put me on patrol of the Cape Flats instead. I rode with Nico Matthee, a forty-seven-year-old white policeman, and Randall Pieters, a thirty-six-year-old coloured cop. Members of the gun dog squad, they had been tasked with searching the tougher areas of the Cape Flats. They wore the khaki trousers and blue shirts of the Metro police, their badges the colours of South Africa: yellow, orange, white, blue, green and black. Hope in a rainbow. Nico, his large stomach hanging over his trousers, a thick moustache seen on many a man in uniform, was engaged and kindly. Randall, with his silver, tightly cut hair, a face pockmarked with acne scars, at first barely said a word.

I squeezed into the rear seat of their car, two dogs in cages in the back and the rest packed with bulletproof vests, handcuffs, shotguns and first-aid kits. The patrolmen both carried Vektor Z88 9mm pistols, a variation of the Italian Beretta. They needed them for this patrol: we were off to the most violent spread of public housing in this landscape of poverty – Manenberg.

Driving in the streets, I noticed that whites drove most of the cars here; the vans were filled with black South Africans. But then we were in an area where there were no white drivers at all and instead only lonely, litter-blown areas of degradation and poverty. It was a landscape of dirty patches of grass, flapping tarpaulins and endless corrugated roofs spreading on either side. Men in tired blue overalls and woollen caps sat by their doorways, and our police car drove slowly past in the morning sun.

'You get about five to nine living in each shack,' said Nico, nodding to the side. These were the breeding grounds for violence. A place where, he said, bad people all too often were seen being publicly rewarded for living a life of crime.

'They call it affirmative shopping,' said Randall, referring to the theft of property from affluent whites by black and coloured youths.

This was home to members of some of the biggest gangs in the Cape, including the Hard Livings, the Clever Kids and the Numbers. There were dozens of smaller hybrid gangs out there, too. In the early 1990s it was thought at least 130 gangs lived in this part of town, with about 100,000 members.[7] God knows what the number was today, but the police said it was worse now than then.

'They deal in Tik,' said Nico, referring to the locally produced crystal meth, 'or mandrax, marijuana, heroin. But the crystal meth here is the biggest seller. And all the shootings around here are gang-related: it's about competing drug territories. They're armed up – Tauruses, CZ 75s, Glocks. It's mainly 9mm handguns.'

Manenberg, with its 70,000 residents, had fourteen homicides and fifty-six attempted murders in the summer of 2013 alone. This year it was even worse: thirty killings in the first four months. The locals said the police were failing in their job.[8] It even prompted the Western Cape premier to call for the army to be deployed. In the first nine months of 2013 there had been over 1,200 arrests. But the area was far from settled. Randall told me to wind down the windows so I could hear the gunshots, and we entered the roughest public-housing district of the roughest area of Cape Town.

The walls were covered with scrawled gang graffiti: turf marked out like wild cats pissing on trees. The Hard Livings controlled this area – the tag 'HL$' everywhere. The remaining areas were divided between the smaller gangs: the Young Dixie Boys, the Naughty Boys, the Junky Funky Kids – unsaid boundaries on every street. Almost 90 per cent of people living here felt there were parts of Manenberg they could never go.[9]

Nico was talking. 'The coloureds here have no discipline. I'm not a racist but since '94 things have changed, but not for them. The whites used to be in charge. Now it's the blacks in charge. The coloureds are stuck in the middle. Sometimes they want to act like whites, sometimes like blacks. This is the problem.'

The Numbers gang, the 27, 28 and 29, were the worst, he said. A four-year-old had been raped and burned by some of them. Then he described how Numbers members were thought to be behind the recent shooting of two cops near Manenberg.[10] A tattooed youth had fired into the prostrate officers as they lay in the dust, shooting one of them through the head. He had taken the firearm from the other. This was a high accolade for gang members here: to become a cop killer. Better still if he got a gun.[11] I looked out of the window and wondered if the police made me more of a target.

Our police car slowly turned and shifted gear, and we were deep into the heart of the district. This dirty grass and stone and concrete land was a scar on the conscience of South Africa. Unemployment in Manenberg was as high as 66 per cent – poverty clung to the houses like mould.

There was a click. Randall slotted a round into his shotgun.

The roads were layered with rubbish, kids playing listlessly. The homes here looked the same as prison compounds or military barracks – straight lines of municipal bare concrete infused with hopelessness. The only colour was the graffiti. Boys stood at a faded corner and stared at us, sullen-eyed.

'The gangs use these young children to hide the guns – they are the runners,' said Nico. 'And the watchers.' Everyone knew everyone here; strangers were noted. Of the 200 people killed in gang violence in Cape Town in the last year, over two-thirds knew their killer.

A dozen warrants had been issued for known gang members in these streets, and there were two more to be picked up. That was why we were here. One of those was wanted in connection with three firearms linked to a specific murder. Then a call came over the radio: the suspect had been seen. He was wearing a white T-shirt, and the crackling voice said he had a gun on him. Nick's foot pulsed on the accelerator.

He drove a hard left, then right, and we hit the pavement and spilled out, chasing breathlessly down a side road into a labyrinth of unnamed streets. Two other armed police were ahead of me, arms outstretched, pistols out, safety catches off. They called out and pushed towards the copper-brown shacks. An old coloured woman, hair in curlers, appeared above, wide-eyed. Perhaps she was right to be scared. Shootings here were harried, spray-in-the-air acts of madness. In the past the old gang violence was more ritualised – gangs battled at an appointed time at night in open fields beyond the residential perimeter, to prevent injury to innocent residents. Today it was more close your eyes and fire blindly – particularly if you were trapped by some fast-approaching police.

I expected a sudden burst of gunfire, but none came. The runner had disappeared. We went through the ramshackle lanes with their pools of hidden shadows and out onto the next sun-blinded street, and there was nobody there except one man wearing a red-stained T-shirt. He was bleeding heavily, because someone had struck him full in the mouth. His neck and chest were covered in tattoos, markings of the gang number 28. He walked past the police and did not stop. His bleeding was his concern, not theirs.

The police were never going to be seen as friends in this area. Over 40 per cent of the people here thought they took protection money from gangsters. And over 80 per cent said the police would not be able to protect them if they wanted to be a witness in a murder trial.[12] So why would a gang member go to them over a split lip?

There was a certain truth in the wounded man's attitude. Later, researching the South African police, I learned that in every month in this country's first seven years of democracy there was an average of thirty-six deaths 'as a result of police action' – 91 per cent of those being shot.[13] The cops here had an image problem.

The gun dogs were out of the cars, barking, but the moment had been lost. The children that rested on each street corner had long ago spread the news that the law was around; those who were sought were gone.

Nico's hair was matted with sweat, and you couldn't tell if it was the heat or fear that had done it. The operation was drawing to a close; the sun had reached its zenith, and it would be hours before a criminal with a gun would venture out to reclaim the streets and the night. The gun units would have to return another day, and my press embed was up.

Besides, the dogs were tired.

Police forces in the states of California, New York and Florida had all refused my request to observe a US Special Weapons and Tactics – SWAT – team raid. Many others never returned my call. So the email from Nevada was not a surprise.

'It is against our policy publicize our tactics [sic.] and our police department does not allow ride-alongs on raids due to privacy issues. We will be unable to accommodate your request,' wrote Larry Hadfield of the LVMPD from Las Vegas's Martin Luther King Boulevard.

This was frustrating, because I knew there were about 50,000 SWAT raids in the US every year, and their refusal felt like another slap in the face. So the email later that day from Chris, a police sharpshooter I had been talking to, was a good one. 'Happy to meet,' he wrote. Then he added that phrase that makes your heart sink: 'but it will have to be off record'.

It's always like this. Soldiers and policemen are inevitably cagey when it comes to speaking to journalists; when they agree to meet they think it's a Deep Throat exposé. So on the one hand you have saccharine police embeds, on the other you have a culture of silence. It's no wonder the police often get away with so much. But Chris was my small entry into what had become a massive issue in the US: the ongoing militarisation of their boys in blue. So we met.

He was one of four sharpshooters in a twenty-one-member SWAT team in a mid-size midwest town and had been in that job for about six years. He had a good lifestyle, earning just over $70,000 in a county where the average salary was $47,000. And because he had to do regular fitness tests – a timed 1.5-mile run, sit-ups, push-ups, bench presses – he was fit. The polo shirt he was wearing was tight-fitting, and his arms filled the sleeves. But his eyes were dull, and it did not take me long to realise that this man was going to give very little away.

'We shoot every other week for about four hours each session, up to 1,000 yards down range,' he said. 'We train to put down a target with a single shot. We aim for either a headshot or the chest area.' In the situations he would find himself in, you don't try to wound. One survey of American police sharpshooters found that 80 per cent of all recorded incidents were fatal – about half hit the head.[14] It takes a certain skill to be able to do that.

'So what sets you apart from other officers?' I asked.

'The single difference between me and my SWAT team members is patience,' he said. It means he often has to spend hours doing surveillance. 'Most of the SWAT guys want to get in and do something – "go, go go" – but my role is more monitoring, watching, calling in about tactical concerns, blocking the path of entry and exit of the criminal.'

He looked a little disconsolate. 'I guess it is not as fun, not as exciting and adrenaline crazy.'

I asked him if he had any regrets, and he said he had used his rifle only the once to lethal effect.

'The guy had beaten up his wife and had a history of mental problems.'

He didn't want to elaborate, but five team members had fired at the same time.

'It makes you think though . . .' he said, with the trace of a southern twang, 'was the guy mentally ill? Could he have been talked out of it before he came out with a gun?' But he had never lost sleep over it and he loved his job.

'I stop the bad guys from getting to hurt the good guys,' he said. It was a simple and uncomplicated way of viewing things.

Yet I was disturbed by what he had told me about that mentally ill man, and the more I read about the world of the police sniper, the more disturbed I became. There were stories of sixteen-year-olds, distraught at bad exam marks, threatening to take their own lives with the family shotgun, then being killed by over-zealous police sharpshooters.[15] Tales of police snipers in camouflage suits being sent into the home of paranoid men in crisis, schizophrenics threatening to cut their wrists.[16]

It struck me this was a world where the gun could easily cause situations to escalate: the inevitable police armed response in the US transforming events into things far worse. And the more I dug, the more I realised how bad the situation had become.

In May 2010 a team of six armed police in Nassau County, New York, was granted a warrant for a no-knock entry to a home. They were after someone in that Long Island house who, their informant had told them, was selling drugs. Their intelligence, however, didn't say that the address they were raiding was two apartments, not one. So when the policemen battered down the door of the downstairs flat and charged in with rifles, voices puncturing the dawn silence, they were faced with an impassable staircase.

'Alternate breach!' they screamed and then rushed outside to knock down the door of the other flat, high on the excitement and the drama.

Iyanna Davis, a twenty-two-year-old, had been woken sharply in the early hours of that May day. Confused, she had hidden in her closet, the violent commotion in the downstairs unit terrifying her. She did not know the intruders were police officers and assumed her home was under attack by thugs. She certainly did not see two of New York State's finest, Michael Capobianco and Carl Campbell, enter her home dressed like soldiers of death – in heavy black boots, thick black combat trousers and black helmets. They were carrying semi-automatics.

The police were later to tell conflicting tales of what happened next. The first story was that Iyanna had leaped from the closet, causing them to open fire. The second was that Iyanna had held the closet door shut against their attempts to open it, causing officer Capobianco to fall down and his rifle to go off. Either way, Iyanna was shot – a single bullet that hit her in the breast and ricocheted through her body, piercing her abdomen and both thighs.

She later told her lawyer: 'I told them I was afraid and do not shoot me, and one officer screamed at me to put my hands above my head. That's when I heard the shot.'[17]

Iyanna had nothing to do with the alleged offence and posed no threat to the policemen, yet she was almost killed by one of them.

Despite this, Nassau County Police Department cleared its officers of any wrongdoing. They agreed to pay Iyanna $650,000, but as part of the settlement the internal investigation was officially sealed. According to Iyanna's lawyers, this was to stop lies told by police officers from coming to light.[18] The policeman who shot her, Sergeant Michael Capobianco, had a salary of over $143,000.[19] He was not publicly reprimanded.

This was not the first time Nassau County police had been involved in the inappropriate use of deadly force. In recent years one of them had shot an unarmed man in the back. Another officer – after a night out drinking – had shot a defenceless cab driver. One more killed a hostage while trying to shoot her armed attacker.

In each case, department investigators reviewed the use of deadly force and, within a day, concluded that the officer's actions were justified. Since 2006, Nassau's investigators have decided that every single one of their officers involved in shooting and killing someone was justified in so doing, this despite the fact that Nassau County Police have been involved in thirty-six shooting incidents in the past four years – a 260 per cent rise from the four years before that.[20] Scrutinising Nassau County Police shootings, though, is hard. They lie behind what's been called a 'thick blue curtain' – the Police Department's refusal to release court documents on deadly incidents.

It is not just Nassau County. US police forces nationwide have been criticised for their overuse of force, for evolving into what's been described as 'warrior cops'. And the most visible element of this is the rapid rise in SWAT teams employed by US police forces.

SWAT teams are, frankly, ubiquitous in the US. There are 56 SWAT teams in the FBI alone.[21] The University of North Carolina at Charlotte has its own SWAT team, armed with MP-15 rifles, M&P .40 calibre pistols and shotguns.[22] And this is just the tip of an iceberg: 13 per cent of US towns between 25,000 and 50,000 people had a SWAT team in 1983. By 2005, 80 per cent of them had one.[23]

Tactics once used in situations like hostage-takings or bank raids are now increasingly being used in everyday police work. Like when in 2010 in New Haven, Connecticut, a SWAT team was sent into a bar because it was suspected of serving under-age drinkers.[24] Or when in Atlanta a SWAT team raided a gay bar where police suspected people were having sex openly – only for a federal investigation to later conclude that the police had lied about such allegations.[25] Armed units were even used to raid barbershops in Orlando, Florida; thirty-seven people were arrested for 'barbering without a license'.[26]

Some of these raids, naturally, had terrible consequences. You don't send in armed men again and again without expecting screw-ups. Radley Balko, author of the *Rise of the Warrior Cop*, has listed over fifty cases of innocent people who have died as a result of botched police raids. There are horror stories of seven-year-olds being shot in the head after police entered the wrong home;[27] of a grandmother shot when her attempt to protect her grandchild was mistaken for something else;[28] of concussion grenades being accidentally thrown into children's playpens.[29]

It also endangers the police. Like when in 2008, after an informant told Virginia police that Ryan Frederick was growing dope at his home, a SWAT team was dispatched to a no-knock raid. Frederick fired his pistol at what he thought was an intruder and killed Detective Jarrod Shivers. It turned out Frederick was a keen Japanese gardener and was growing oriental maple trees, not marijuana. If the police had entered with a warrant, the misunderstanding would have been resolved peacefully, an officer would not have died, and Frederick would not be now serving ten years for manslaughter. But they did not.

The lessons learned from this, for me, came in the form of a hard fact. More police officers were shot and killed in the US in 2014 than were shot and killed in the last fifty years in Great Britain.[30] This suggested to me that police officers, with their ready use of guns, may well be escalating situations and, in so doing, putting themselves in avoidable danger.

But little seems to be being done to address this overuse of force. The problem is that, in a post-9/11 world, the US police have become what Balko calls 'a protected class'. Few politicians want to oppose them, so they are rarely held to account successfully, and, despite a growing media focus on their actions, no one seems to be effectively restricting their powers. Instead, America just keeps arming its law enforcement officers with military-grade equipment.

It was something that was glaringly obvious on the streets of Ferguson, Missouri, in the summer of 2014. When Michael Brown, an unarmed black teenager, was shot and killed on August 9 by Darren Wilson, a white police officer, protests riled the area for weeks. The world looked on as the streets of an American town seemed to descend into a war zone – masked police with tear gas, beanbag rounds, flash grenades and rubber bullets descended on Ferguson. A policeman was filmed saying, 'Bring it, you fucking animals, bring it', and the issue of the militarisation of America's police became the focus of endless media columns and articles.

From an outsiders' perspective it seems a certain madness has descended on American law enforcement. Between 9/11 and 2013 the Department of Homeland Security handed out $34 billion in 'terrorism grants' to local police forces to fund counter-terrorism efforts.[31] This, alongside rules on civil-asset forfeiture allowing the police to seize anything they reasonably consider the proceeds of crime, means that, despite America's failing economy, there is still a huge amount of money available for the police and other agencies to spend on things like guns.[32] And spend they do.

In 2010, the US Bureau of Alcohol, Tobacco, Firearms, and Explosives awarded two contracts worth $80 million to Glock and Smith & Wesson for the supply of .40 calibre service sidearms. In 2013, the Department of Homeland Security put out a tender to buy 1.6 billion rounds of ammunition over the following five years.[33]

Such wilful and expansive spending is best summed up by discrete examples of security exuberance. In Augusta, Maine, a sleepy place of less than 20,000 souls where no officer has been shot and killed in the line of duty since the Statue of Liberty was erected, police bought eight tactical vests for $12,000. Des Moines in Iowa bought two bomb-disposal robots at the cost of some $360,000. Richland County, South Carolina, purchased a 'Peacemaker' – an M113A1 armoured personnel carrier that has a belt-fed, turret machine-gun that fires .50 calibre rounds, big enough to shoot through concrete walls. And throughout the US the old police six-shot revolver has been largely, and expensively, traded in for semi-automatic pistols that carry up to eighteen rounds. This mass purchase of handguns has been done in the face of evidence that in most shootings involving the police no more than three shots are fired.

Of course, many would say the American police are entirely justified in arming themselves to the teeth – look at what they have to contend with. ShotSpotter, a company that specialises in gunfire detection through urban listening monitors, looked at the data from forty-eight American cities over 2013. The company found that fewer than one in five gunfire incidents were reported to the police. In some neighbourhoods less than 10 per cent of gunfire was reported.[34]

Even with so many illegal shootings, the hyper-militarisation of America's police force is still of concern. In 2012, according to FBI data, 410 Americans were 'justifiably' killed by police – 409 with guns.[35] There is the very real issue of a police-criminal arms race: that criminals will acquire similar powerful weapons to combat the police.[36] It ignores the National Institute for Justice review which concluded that assault weapons were 'rarely used in gun crimes' and that high-powered weaponry was hardly ever used in the killing of police officers.[37] It also ignores the fact that the majority of raids conducted by SWAT teams turn up no weapons at all.[38]

This weaponised and hostile face of American law enforcement, moreover, results in a hardening of stances, where community policing is lost to heavy-handed brutalism. It is a sort of 'crush them' approach to criminals that is reflected in the fact the US has the most prisoners of any nation in the world – over 2.2 million men and women behind bars, almost one in every 100 adult Americans – and a philosophical approach to order that results in situations such as kids being charged with a felony for throwing peanuts at a bus driver,[39] or schools handcuffing children for petty things as trivial as not wearing a belt.[40] Such logic has meant that the US today is a nation where there are more security guards than teachers.[41]

For me, though, the most compelling reason to be wary of not properly scrutinising each and every SWAT team deployment and every bullet fired is the danger that it gives rise to police impunity. And this impunity can often rapidly lead to an ugly reality: that of extra-judicial murder.

It was the death of an activist that had brought me to the southern Philippines in the early summer of 2008. I was reporting for ITN, covering a story about a rise in police brutality throughout this South-east Asian island nation. This man's tombstone was the heavy cross upon which we could hang the story of countless other sad deaths.

The broad white coffin was large for such a slight man, and as it was lowered into the ground, the anguished lamentations of the mourners drowned out the sound of the tropical downpour beyond.

The person being buried was Celso Pojas. He had been a political leader, the secretary general of the Farmer's Union in Mindanao, and it had been a bullet from a paramilitary death squad that had placed him in the sodden ground.

A few days before, forty-five-year-old Celso had been savouring a coffee in his office in Davao City, when he got up. 'I have to get a few cigarette sticks,' he said and walked outside. These were his last words. Moments later his colleagues heard the staccato drum of gunfire and rushed out to see their friend face down, dying. He had been getting death threats for months, the latest delivered on the evening before a series of transport strikes he had set up. The message was unequivocal – don't meddle in politics.

Celso's death followed a pattern. The killers had arrived on a motorcycle that did not have a licence plate. They wore baseball caps and buttoned-up shirts, their firearms tucked into their waistbands. They favoured .45-calibre handguns, weapons commonly used by the police and prohibitively expensive for criminals. They shot without warning. And as quickly as they arrived, they left.

No one knew who had killed Celso, but the general consensus was that there had been police involvement. Government officials and members of the police here in Davao had been implicated in twenty-eight killings, mainly between 2007 and 2008.[42] A further 298 killings had been carried out since January 2007 by the Tagum City Death Squad.[43] Such murders had caused the charity Human Rights Watch to conclude that the government had 'largely turned a blind eye to the killing spree in Davao City and elsewhere. The Philippine National Police have not sought to confront the problem.' These might seem strong words, but words far stronger came from the mayor of Davao City, Rodrigo Duterte, when he said: 'If you are doing an illegal activity in my city, if you are a criminal or part of a syndicate that preys on the innocent people of the city, for as long as I am the mayor, you are a legitimate target of assassination.'[44] *Time* magazine was to call Duterte 'the Punisher'.[45]

Of course, some saw Duterte's firm tactics as justifiable, claiming that he had brought a level of security to this violent, sweltering city that it had not seen for years. But what happens – I wanted to know – if, in searching for peace, you created desolation?

Celso's mourners had this question in mind when they met before the funeral at Davao's Freedom Park. Muscular farmers who had travelled 160 kilometres from Compostela Valley, plantation workers in thin cotton shirts bussed in 270 kilometres from Davao Oriental. They lined the streets of Araullo, Quirino and Ponciano and filed in solemn anger past his open coffin. What good, their faces said, is a peaceful city if you live in fear of those you entrust with keeping such peace?

'These death squads,' I was told by an activist who refused to be named for fear of retribution, 'were run by police officers. They give them weapons, ammo, bikes. They pay $1,000 for each killing, if that.'

Police killings like this one are among the ugliest faces of gun violence I have seen. And when I was later to sit and interview Celso's family after the service, the impunity, the helplessness of it all made the room mute with anger. The women sat to one side, dissolving in tears. The men stood silently in their loose tops and flip-flops, resting up against the moist walls of Celso's modest home. Someone handed me a glass of warm fizzy orange and a chocolate snack, and

I wondered if Celso had bought this bottle of pop or this sweet to give himself a treat.

'Why?' the father said. 'Why my son? He was so good, he helped us all.'

As a journalist, there is not much more you can do but write up such things. In cases like this, where corruption sinks so deep you can't see its end, you can hardly hope for an arrest or a review. Or even an interview. But everyone in that close room was affected and not just by the loss of their loved one. They had seen a destruction of trust, something that can never be truly regained. And it was an abuse of power that was certainly not confined to the Philippines.

Every year, Brazilian police are reported to be responsible for at least 2,000 deaths nationwide – an average of five people a day.[46] The victims are often recorded as having been 'killed while resisting arrest'. And then you read that one policeman killed sixty-two people and registered each of their deaths as such, and it all becomes a bitter joke, not even an excuse.[47]

Elsewhere it is just as bad. In India, reported incidents of police firing on civilians have almost doubled in the last decade.[48] Many are called 'encounter killings', confrontations used to justify extra-judicial murders – implicitly seen as an acceptable response to crime or terrorism.

In Jamaica it is said that one of every two police officers who spends twenty-five years on active duty will kill in the line of duty.[49]

The worrying thing is that such brutal police tactics are often seen as the only way to clean up the streets. The idea of a noble cop taking the law into his own hands is the stuff of countless Hollywood movies. But the abuse of power, the arbitrary use of capital punishment, the absence of a fair trial and the risk the police might accidentally kill an innocent – all of these mean that the moment a cop purposefully takes a life he ceases to be a policeman and becomes a killer. And when your police are killers, there's really not much room for hope.

8. THE MILITARY

The tragedies of war – Iraq – travelling to the bloody circus in 2004 – visiting the Tree of Knowledge – getting shot at – madness and violence unfurling – Israel's violent past reflected in guns – tea with an unusual sniper – a visit to a Jewish anti-terror training camp – Palestine's tragedy – a wounded boy, a grieving father – Liberia's past visited – child soldiers and adult tales

Militaries and guns are synonymous; an army that is not armed isn't really an army. It's no surprise there are 200 million guns in the hands of armed forces around the world: about one in five of all guns. So you can't write a book about firearms without understanding the gun's role in war and its military use – for good or bad – in defending a nation's sovereignty. After all, there are only fifteen countries in the world that do not have a military, and six countries that have militaries but no standing army.[1]

In the remaining countries, the 200 million weapons are unevenly distributed: the armies of just two states, China and Russia, have almost 25 per cent of them. And they are certainly not all in use. Around 76 million guns in the hands of armed forces lie idle; they are deemed 'surplus' – stockpiles make up about 38 per cent of all military small arms.[2]

But when war does break out, it is clear that there are enough guns out there to cause untold carnage. Since the end of the Second World War there have been over 2,100 conflicts in more than 150 locations.[3] In 1998 a charity called Project Ploughshares said that in over 'three dozen current wars, probably 90 per cent of killings are by small arms . . . in the past decade alone they have caused more than 3 million deaths'.[4] This is a bold statement and one that led some to say that about 300,000 people were dying from guns fired in conflicts every year.

But charities have reasons to sound the death knell a little too loudly, and you can't believe every fact you are told, not least because, in this case, the Geneva Declaration, a diplomatic initiative endorsed by 100 countries, has a much lower figure for those killed by all weapons of violence – bombs and guns included. It estimates that, of over half a million people killed globally by armed violence every year, only 10 per cent of violent deaths, about 55,000, happen in conflict or a terrorist attack. An even lower number, then, would have been killed specifically by guns.[5]

Of course, these statistics don't include the numbers of those injured by guns in wars. This should concern us because, as I had seen in South Africa, it is clear that there have been marked improvements in trauma surgery. So the death toll today in some wars might be lower than it would have been years ago, but this does not mean that wars are getting less violent. We are just better at fixing people.

What we are pretty clear about, though, is that the role of the gun differs markedly from war to war. The AK47's popularity in the Republic of Congo meant over 93 per cent of deaths were from gunshots there.[6] We know thousands of civilians were killed by guns in Iraq – all too often the result of kidnappings and assassinations. But in Uganda, the conflict waged by the Lords Resistance Army – a militant cult led by Joseph Kony, which seeks to establish a theocratic state based on the Ten Commandments – shows that knives and clubs are also frequently used to murder and terrorise.

The widespread military use of explosive weapons has also had an impact. In Cambodia in the mid 1990s and in Thailand in 1980 more civilians were killed by mines than by guns. And, at the other end of the spectrum, the sheer quantity of air-dropped bombs in Lebanon in 2006 meant that less than 1 per cent of people were killed in that conflict from gunshots.[7] In general, though, it's estimated that guns account for between 60 and 90 per cent of all direct casualties in war – a heavy toll however you look at it.

The harm guns cause has also changed over time. In the American Civil War, guns accounted for about 75 per cent of combat casualties. By the Second World War only about 18 per cent of military casualties had been shot. This shift is down to a few things. The nature of warfare has changed over time: explosive weapons are much more likely to be used now than they once were, and this pushes down the proportion of those injured by gunshot. Despite criticism that US soldiers in Iraq and Afghanistan were given inadequate or even useless equipment, the general truth is that, today, soldiers are much better protected: improvements in bulletproof vests, armoured personnel carriers and helmets mean getting shot is now less likely to kill you. And soldiers are increasingly taking on targets from miles away – as the use of drones makes clear – further reducing their chances of being caught in the crossfire.

If a soldier from a relatively developed military force is unlucky enough to be shot, the swiftness of getting treatment to them has massively improved their chances of getting off the field of battle alive. Gunshot lethality prior to Operation Iraqi Freedom was about 33 per cent. Now, according to US military data, it's less than 5 per cent. The only thing that has not changed over time is the lethality of headshots.[8]

The impact of guns in war also changes during the course of each conflict, and not only because surgeons are getting better at what they do. At the start of the Russian involvement in Afghanistan in 1980 about two-thirds of conflict casualties were from gunfire. By the end of the decade the Russians had learned the hard way about the skilled marksmanship of the mujahedeen, and so they kept their heads down. By 1990 only 28 per cent were from gunshots.

All of this shows one thing: that the role of the gun in soldiers' lives has changed as much as the nature of war itself. As part of my work, I have walked beside troops around the world and I have seen that each military deployment is unique. From filming the menacing metal bristle of the borders of South and North Korea to watching British squaddies walk in stern silence in the soft fields of Kosovo,

I have just one simple observation: for most soldiers, in peacetime the gun is just a thing, something they carry with them, something they oil and clean, eat with, shit with, even sleep with, not something that they really talk about, unless they lose it, or a screaming sergeant makes them run with it over their heads for an hour. But in war the relationship between the soldier and the gun changes completely. This is why, in order to understand how the world of guns impacts militaries, you have to travel to war itself.

April 2004 – I was in Basra, working for the BBC with my reporter, Sam Poling, one of those journalists who chases at stories with a ferocity that you can only admire. She and I were embedded with the Argyll and Sutherland Highlanders, a Scottish regiment that had fought in Korea and in Aden, in the Boer War and in the fields of Flanders, and now it was raising its colours here in southern Iraq.

We had just seen the Tree of Knowledge – a broken tree in a broken land, neglected and unloved, but the infamous tree all the same. It stood in the centre of Al-Qurnah, a small, windswept town about 70 kilometres north-west of Basra. This jujube tree was near the confluence point of the Tigris and Euphrates rivers – where they joined to form the Shatt al-Arab. It was the place where Iraqis claimed Eve had plucked that forbidden apple and with her first bite allowed knowledge to ruin paradise.

On the day we went there the tree stood dying, plastic bags eddying around it, and boys with snot on their faces and holes in their trousers kicked the dust at its roots. But they only did this during the day, because no one walked the streets here at night, except killers. After all, this was Iraq, and this was war.

We turned and began the long drive back to the British army base in silence, as no one likes to see paradise lost.

Then came the gunshot: a stark, blue staccato snap and a screech of brakes and a tumbling out of the Land Rover. Out we ran, onto the sandy banks of the road, over the pebbles and plastic that littered the sides of the highway, and, breathless, we landed in a gully.

'They shot at us! They fucking shot at us!' screamed one of the British soldiers. They tensed and raised their rifles, but the car was already speeding away.

We had been due to return to Britain that day, but the road to the airport was too dangerous. The army had already lost a few soldiers on the way to that baked tarmac strip, and with the threat as high as it was, the colonel said we would just have to wait. Something bitter, disconcerting and violent was happening.

On the flight coming in I had been strapped into a stand-up harness at the front of a Hercules troop carrier. We were flying down low straight from Cyprus, and the silver line of the Qamat Ali canal glinted beneath us in the night; on either side stretched the silhouette of ancient desert lands. We were flying blacked out, a dark speck in a dark sky, but then a red button flared upon the pilot's dashboard, and the crackle of a command came into the headset.

'Incoming. Release one. Release two.' A ground-to-air missile was fast approaching, and the pilot fired off decoy flares, lights spinning behind us into the pitch-black. Our plane tilted sharp to the right, and the threat passed as fast as it had come. But it was clear we had begun our descent into something.

The ragheads were fuckin' losing it, the soldiers had said to us later, and with swagger. There was excitement in their voices at the prospect of having a decent 'contact' to tell their mates about back home, but then you saw the youth in their eyes. They yearned for what they should have feared.

If I was honest, I too was glad we had been shot at in such a neat little way, because journalists can't go to a war zone and not wish to see a gun in action, no matter how much they wrap it in platitudes. That's why you are there: not to film soldiers grumbling about their fat-saturated dinners or how much they miss their mothers, but to film the coursing adrenaline rush of wide-eyed terror and the sharp crack of a gun's retort.

This is why the Iraq War, above all the others, defines my view of the military use of the gun. I had been to many conflict zones – Somalia, Pakistan, Colombia, Nagorno-Karabakh – places marked as much by war as peace. But Iraq was different. The war there, perhaps because tensions were so high and because I was unarmed and yet everywhere surrounded by guns, was compelling and vivid and unlike anything I had felt before. Guns here defined life; they were the only things that mattered.

The Americans had begun the now infamous battle of Fallujah on the day we landed: a huge assault to clear a city far to the north-west of us. Their targets were those responsible for the gruesome killing of four US Blackwater military contractors. Stumps of burned American flesh had been dragged through the streets, captured on a thousand broadcasts. And the US, as the US does best, retaliated with heavy force.

On the night of 4 April 2004, forces under Lieutenant-General James T. Conway launched a major assault to 're-establish security in Fallujah', circling it with 2,000 troops. The subsequent violence had shaken Iraq, sending ripples through the country as far south as Basra.

So what started as an ordinary media embed with a British infantry regiment changed into something much more dangerous, as a slow unfurling of anger and blood-revenge gripped the city. We had had bricks thrown at us by baying mobs, seen army Land Rovers transformed into black, burned-out skeletons on the lawless roads. Sam had even been attacked by one of the Iraqi soldiers the British were training. And now, in a ditch beside a village we had never heard of, we were being shot at.

But the danger passed as quickly as it arrived, so we dusted ourselves off and got back into the Land Rovers and drove slowly to Camp Riverside – the tidy army outpost that was about 25 kilometres away from Basra. The camp was, at the height of the Ba'ath party's reign, the summer home of Ali Hassan al-Majid, the Iraqi defence minister and chief of the Intelligence Service. Al-Majid was more commonly known as 'Chemical Ali' for his use of chemical weapons in Iraq's attacks against the Kurds, and his old home was graced with a beautiful view over swaying fields and bulrush-lined canals. Rumour had it that it was once filled with whisky sours and bubbling Jacuzzis and unspoken exploits; thoughts that would set any lonely British soldier's mind aflame.

But just as it was a place for secret and illicit pleasures, so too was it a place of protection from the bloody circus that was grimly being performed across Iraq. There were now high walls and watch-towers and sandbags aplenty here, and we slowly zigzagged through the protective escarpments designed to stop suicide trucks. The high metal doors of the camp were pushed open, and we were back at base. The men disembarked and walked without words to a small rectangle of sandbags. There they unclipped their magazines, making their weapons safe. They did so with easy, fluid movements, as had been drilled into them.

We headed over to the cookhouse on the far edge of the camp, the smell of frying food in the air. I made a beeline for one of the sentries defending the outer reaches and ducked into his neat sentry post, with its swollen sandbags. He smiled and leaned into his weapon, an L7 general-purpose machine-gun.[9] Neither of us spoke. We looked to the west.

A boat slowly drifted down the waterway. On the deck, a dark-skinned man in a white *dishdasha* stood, his hands resting behind his back. Our gazes met, and the boatman's eyes narrowed; you couldn't help but see menace in that furrowing of brows. And the machine-gunner traced the boat's slow passage downriver until the man passed around a bend and out of sight. The soldier let loose a long line of spit and winked. I turned to go, knowing what that wink meant. We were inside this camp, they were outside the camp and, until we knew differently, each and every one of them was a killer. And the oiled and cared-for machine-gun, with its ability to let loose ten rounds in a second, was a thing of protection for all of us.

The next day it got a lot worse.

We awoke before the sun and headed south as the city stirred. The cars drove, engines bull-loud, into Basra city and the platoon's lieutenant, a young Scot from the Borders who one day talked about returning from these dry lands to help run a family fishery, shouted through the grill that separated the front and the rear of the Land

Rover. Things were heating up, he said. Two of the regiment's convoys had just been hit with rocket-propelled grenades.

This was just the start of it. By the end of that day there had been over fifteen contacts in the city – exchanges of gunfire between British soldiers and militants. We wanted to get out to see how the day would unfold, but the colonel of the regiment said it was just too dangerous, so we waited in the main base, and each time a convoy came back in, men would tumble from their armoured vehicles, their uniforms flecked with blood, their eyes wild. They spoke in fast, breathless clichés, resting on old sayings to try to explain what had just happened.

'It was hell,' one said and told how his colleague had his foot blown clean off by a rocket grenade. Another had been shot in the hand. One even told how his grandfather had given him an antique coin, the old man having picked it up in these same desert lands in the Second World War. His warrior grandson had kept it in his top pocket over his heart and forgot all about it. Then his convoy was hit in a rocket attack and it was only back at the HQ, when he pulled out the coin, that he found a piece of shrapnel embedded in it. 'That coin saved my life,' he said.

The words spilled out from their fresh, youthful mouths, the adrenaline keen in their faces, and they told how they had let rip with their weapons, facing ambush after ambush, because this was a war that, in many ways, was being fought with the age-old technology of guns. They had become crucial weapons in this urban guerilla conflict, a place where planes or tanks or mortars were sometimes too blunt weapons of violence to use.[10]

The figures testify to this. It's been estimated that 250,000 bullets were fired in Iraq for every insurgent killed by the US military.[11] That works out at about 6 billion bullets being shot by US forces between 2002 and 2005.[12] Indeed, so many bullets were fired by American soldiers that the three ammunition contractors who supplied ammunition to the US military had to spend almost $100 million in upgrades just to keep up with demand.[13] Even this was not enough. Over 300 million rounds were bought from commercial companies, including Israel Military Industries and Olin-Winchester.[14]

Someone, somewhere, was making a handsome profit from the

US and British incursion into Iraq. It's no surprise that Whitehall figures put the cost of British funding of the Iraq conflict at $13.7 billion,[15] or that, between 2005 and 2008, the annual cost for each American soldier there rose from $490,000 to $800,000.[16]

Such profits should be weighed against the numbers of lives lost in Iraq. Of the estimated 122,843 civilians killed there between 2003 and 2014, about 55 per cent died from gunfire.[17]

Years after leaving that military embed, when I was working on Wikileaks' Iraq War Logs, we found that more than 80 per cent of people shot and killed in incidents at US and coalition checkpoints were civilians. Over 681 innocents died – at least thirty of them children – compared to just 120 Iraqi insurgents.[18]

There were other, equally terrible times when the sound of murderous gunfire penetrated the fog of war and caught the world's attention. Like when, in November 2005, a group of Marines went on a shooting spree and killed twenty-four Iraqi civilians.[19] Or when, in 2006, US soldier Steven Dale Green raped a fourteen-year-old Iraqi girl after killing her parents and younger sister and then shot the sobbing teenager in the head.[20] Some hear these stories and say: this is war, shit happens. But these incidents should be remembered, because while nations have a right to self-defence – the argument put forward by the US and the UK – they also have a legal and moral responsibility to adhere to international humanitarian law. They have a duty to ensure arms are used appropriately and proportionately. And in Iraq they, clearly, were often not.

The tragedy of Iraq, though, is that there has been no proper attempt at investigating war crimes committed by Coalition soldiers on duty; even politicians who opposed the war and are now in power seem reluctant to open that Pandora's box. This political silence means that, largely, the horrors of what guns do to people in war remain unaccounted for and unseen. And future politicians may all too easily forget what war means when they next send young men into the field of battle, there to face all of its iron indignation.

Perhaps this is not a surprise. Iraq was fought in the way that guerilla wars are so often fought: in the shadows, largely away from the glare of the media. On that bloody day in Basra, it got so bad that the whole camp went into lockdown, and all patrols were suspended. The army decided it was simply too dangerous to head out of the gates because of the attacks and the rioting mobs.

As the sun rose higher in the sky, and the sound of gunfire resounded in the distance, Sam and I sat down to talk. I suggested that we just get a cab to where the maddening conflict was unfolding. Sam, being the journalist she is, felt the same. Captain Johnny though, the officer in charge of the unit we were following, took us aside and told us it was a suicide mission, that the Quick Reaction Force would not come to get us if we got into trouble. Even the Iraq Civil Defence Corps colonel told us not to go. But I wanted to go, and badly. I wished to catalogue the violence, to witness what Britain's role in Iraq had become – so we ignored their pleas. We packed up our kit and got the unit's interpreter to call his uncle who had a cab, and headed out the gate.

As we waited, though, something felt wrong. The taxi had yet to come, and in the slow minutes that followed, my bravura turned to doubt and doubt to fear. I turned to Sam. 'This is indeed a suicide mission.' She nodded. We packed up the filming kit and walked back through the gates. Without the protection previously offered by British army guns, we had to leave the day unrecorded; we had little desire to film our own deaths.

Parts of this war, as with all wars, had to be reduced to silence. When the gun's impact is so extreme, nothing will properly describe or explain it. And if you were there to describe it, then you risked too much. Looking back on those days, it feels futile for me to try to explain at all the role of the gun in that war – at least from those few grasped experiences. Just as witnessing a dozen wars would lead me to a dozen different conclusions about the role of the gun in each and every one.

Faced with such complexity, I wanted instead to focus on a place where the gun's role was more unequivocal. So I chose to travel to a country that for the last twenty years has been repeatedly called the world's most militarised nation.[21] A place born from gunfire and existing by gunfire: the state of Israel.

The ancient port of Jaffa rose like a dusty tombstone to the south, and the sun bleached the Middle Eastern sky. It was furiously hot, and I was poorly dressed for it. My heavy boots and shirt were draining me, but it was not the heat that caught me off guard. What surprised me were the questions, because I've never been to a museum before and been asked either for my passport or whether I was carrying a machine-gun. But that is what a uniformed youth wanted to know as I tried to enter the Israeli Defence Museum, a short distance from the surfer's beaches of Tel Aviv.

I thought this museum was as good a place as any to start. The Israeli Defence Force, or IDF, had refused me an interview – mainly, it seemed, because I was an investigative journalist. I had been interrogated at Tel Aviv airport for five hours for the same reason. But I could not be prevented from seeing the guns of Israel's history just because I was a pain.

The youth looked long at my stamped visa, a slip of paper stapled into my passport, and nodded for me to go to the ticket office, where another conscript on her two-year service was looking even more bored. It was 15 shekels, about $4, and she ushered me through with a flick of her wrist.

I walked past a turnstile, past a short line of grey, industrial concrete walls and out into a courtyard of focused, broiling air. It was over 95 degrees Fahrenheit, and you sympathised with the guards in their thick uniforms under the unwavering disc of a midday sun. A map pointed me to the huts I wanted – numbers 10 and 14. The first was for 'The Six-day War' exhibit, the other 'The Yom Kippur War'. But I was not drawn there by the histories of Israel's wars, I wanted to go there because these huts were filled with guns.

Perhaps it was no surprise that the IDF museum would dedicate its gun rooms to such influential conflicts in its brief history. After all, the Israeli army is awash with firearms – about 1.75 million of them.[22] This works out at roughly 22 guns held by Israeli soldiers for 100 of its citizens. In Egypt that number is 2 per 100, in Jordan it's 4.[23] Perhaps just as these two historic wars define Israel, so too do its weapons.

Passing heavy artillery exhibits – field guns painted in uniform desert-brown besides small clusters of black shrub-like sculptures of welded-together rifles – I walked to the first of the two exhibits housed in a large Nissan-style building. The Six-day War hut was dedicated to the 1967 conflict between Israel on the one side and Egypt, Jordan and Syria on the other. It was not six days long – it lasted 132 hours and 30 minutes and left Israel with the largest territorial gains of all of its modern conflicts. Sinai and the Gaza Strip came under Israeli control; so too did East Jerusalem and the West Bank, while the Golan Heights shifted from Syrian to Israeli rule. Seven hundred and seventy-seven Israelis lost their lives – their enemies lost over twenty-two times that. The sorrows of modern Israel were firmly laid down within that shortened week, and this hut, filled with cooling air and 608 revolvers and pistols from around the world, embalmed its memory.

The handguns that filled this room were what I had come to see. They lined its walls in white-edged cabinets, each fixed to a perforated screen with small plastic ties. There were Israeli Uzi 9mms, American Desert Eagles, Italian .22mm Berettas and Belgian 7.65 Brownings. Names that conjured up forgotten violent times: Makarovs and Webleys, Mausers and MABs. At the far end, in front of a cabinet of British sea-flares, was a display for bullets, 126 rounds placed in neat rows, each section divided into different types of firing mechanisms: centre fire, pin fire and rim fire rounds.

Then I realised that there were no explanations here, no details given. I had hoped for an insight into the role that guns played in the Holy Land – a sense of why Israel needs so many of them. But it was just a room full of pistols and bullets. Devoid of context, such as I found with the police armoury in São Paulo, or a guide, as given me in Leeds, it was as boring as the sun outside was hot. So it is with many military museums. As if guns themselves deserved veneration: a room where you have to speak in a hushed voice and look solemn. I took a selfie with my phone and wondered just how long it would take someone to run in and shoot me if I broke one of the glass panes to get a pistol. Then I did myself a favour and left.

The next hut, dedicated to the Yom Kippur War, had two green-clad mannequins facing you as you opened the door. For a second I was spooked. But this room was filled with rifles, not people. There were forty cabinets of them, white-framed, glass-fronted as before. Inside were long guns; muzzle-loaders; assault and sub-machine-guns; light, medium and heavy machine-guns; training and target rifles. Guns from all over the world – Chinese, Polish, Egyptian, Lebanese, Greek, Bulgarian.

Beside the door, perhaps in a spirit of revolutionary independence, was a gift from 'Ye Connecticut Gun Guild': a Kentucky Rifle – the one that once struck fear in the hearts of English foot soldiers in the American War of Independence.

Then there was a cabinet dedicated to rifles made by Israel Weapon Industries, including the famous 9mm Uzi sub-machine-gun. Other cabinets showed the first attempts to create rifles here in the Holy Land. Beside them was a section on European homemade guns from the 1940s, and you thought of the small groups of Jews who armed themselves in those terrible anti-Semitic days.

Then I saw a small photo. Five men in black and white – they were in a field, and each held a long rifle. It was the national team for the fifteenth Olympic Games, in Helsinki in 1952. 'The first appearance of Israeli's sharpshooters in an international competition,' said the caption.

This celebration of their skill in shooting was a stark one. The subtext was not their ability to put a hole in a target a hundred yards away, rather, it was the state's ability to end the life of an enemy at ten times that distance, because this room was lined with sniper rifles. And the cabinet that had the most sniper rifles had the Israeli sign over it.

A Lee-Enfield, a Mauser, an M14, and a Galil Sniper Rifle: all laid out. 'In Israeli Defence Force specialist units,' the card next to the Galil read, 'since 1983 and still in use.'

In a way this was what I had come to Israel to see. How the Israeli military, engaged as they were in such a drawn-out conflict, used guns both in self-defence and in attack. And, of these, the sniper's rifle was the most intriguing, because it spoke of a very specific intent – not just of fire and spray, but of targeted death.

Ultimately, though, the guns here were unable to tell me of the stories and deeds that lay behind them, so I headed back into the heavy heat. I needed an interview with someone who deeply understood the role of the guns here. I needed to meet a sniper.

I had not expected it to be a woman. I had asked a local journalist to help me set up the meeting and, in a flippant way, had assumed it would be a measured and quietly violent man. But she was twenty-seven years old and had a light-touched beauty, and she was not what I thought a sniper would look like at all.

We had agreed to meet in the courtyard of the American Colony Hotel, the light Jerusalem-stone-built hotel in the east of the Holy City. The American Colony was set up by Baron Ustinov – grandfather of the late actor Sir Peter Ustinov – when he found the Turkish inns of the time unacceptable and wanted suitable accommodation in Jerusalem to house his visitors from Europe and America. Today, it's a meeting place for journalists, spies and politicians – all drawn to its luxuriant gardens and to each other.

'I've just seen Tony Blair,' the sniper said as she walked past the quiet diners sipping coffee in the shadows of the central courtyard. She was Anglo-Israeli, articulate, well-educated and politically liberal. She defied every preconception I had about snipers.

Her main role had been as a trainer in the Israeli Defence Force. She once ran courses that sought to ensure each infantry unit in Israel had at least one sharpshooter, sometimes more. 'We would teach them how to calculate wind, range, how to deal with their gun malfunctioning; the course was theoretical but it meant going to the range and practising camouflage in urban areas and in the field,' she said.

I imagined she also taught these soldiers to aim for 'the apricot', that small area that lies between the top of the spine and the brain; where a bullet will bring a man down without reflex. They call it a 'flaccid relaxation', and a good trainer would teach you to hit this spot every time.

The sniper training was intense and demanding, she said. Soldiers worked in pairs, the shooter and his observer. The observers were there to analyse the distance, the weather conditions and the wind speed and then give this information to the snipers. The teams endured months of intensive training – carrying weights of up to 60 per cent of their body mass for over 30 kilometres at a time. They developed the ability to make themselves invisible through camouflage; to locate enemy snipers; to pick up on the smallest of details that let them track their target. As one sniper, quoted in the IDF's official blog, said: 'Sometimes you can be focused for two hours and nothing happens. Then, the target comes, and you have two seconds . . . I stand by what I do, though, because these people were aiming at me, my friends and the people of Israel.'[24]

The Israeli army clearly realises the value in training up snipers. Not only were they good value for money (a sign at the US Marine Corps sniper school reads: 'The average rounds expended per kill with the M16 in Vietnam was 50,000. Snipers averaged 1.3 rounds. The cost difference was $2300 v. 27 cents').[25] They were also a psychological weapon.

Certainly, Israel had, over the years, developed a skill at precise killings through the use of helicopters, drones and elite units. Snipers played a major role in this precision warfare, or 'focused foiling' as it is sometimes called. On 14 December 2006, the Supreme Court of Israel ruled that targeted killing was an acceptable form of self-defence against terrorists. How targeted it is, though, is up for debate. According to the Israeli Human Rights organization B'Tselem, these so-called 'targeted' killings took 459 Palestinian lives between September 2000 and June 2014. Of these, 180, about 39 per cent, were civilians or people who did 'not participate in hostilities'.[26]

But this focus on training snipers by the Israeli Defence Force was certainly not a new development in warfare. The verb to snipe first was conceived in the 1770s among British soldiers serving in India. There hunters who proved themselves adept enough to bag the elusive snipe bird earned themselves the title 'sniper'. The British further honed this skill in combat when they formed the Lovat Scouts, a Highland regiment that served during the Second Boer War in 1900.[27] This unit was the first to put on a ghillie suit: a camouflaged outfit designed to resemble heavy foliage.

It was in the Second World War when snipers really seized the attention of the world and became the powerful propaganda tool they are today. These sharpshooters became pin-ups of the battlefront and sent so powerful a message of terror that German snipers were celebrated back home. They were treated to elegant wristwatches after fifty confirmed kills, a hunting rifle following 100 kills, and for 150 kills they were sent on a personal hunting trip with the Reichsführer of the Schutzstaffel (SS), Heinrich Himmler himself.

Some snipers became renowned for their deeds. The most famous was Simo Häyhä who, during the 1939 Winter War between Finland and the Soviet Union, in temperatures that dropped to −40°C, was credited with 505 confirmed kills of Soviet soldiers. All of these kills took place within three months, meaning Simo averaged about five a day. So good was he that he was known as 'White Death', a legend bolstered by such facts as the one that he used the standard iron sights on his Mosin-Nagant rifle rather than telescopic sights, because to use a glass sight you had to lift your head higher and so risked being seen by the enemy.[28]

Of course, Hollywood contributed to the myth and mystery of the sniper. Films like *American Sniper* and *Saving Private Ryan* show hard-bitten American southerners who dispatch the enemy with clinical exactitude. And the 2001 epic *Enemy at the Gates* celebrated the duel between the Russian sniper Vassili Zaitsev and his partly fictional Nazi foe Major Erwin König, as the battle of Stalingrad raged around them.

Such hero worship endures today. Websites like snipercentral.com are packed with details about the tactics employed by snipers, specifications of their rifles and scopes, and league tables of the numbers of kills that famous snipers have chalked up.

Reading these kill lists, you are struck by how prominently modern wars feature in them. This is because as rifle power, sight technology and ballistics have improved, so the distances at which a sniper can hit his mark have lengthened. In 2009 thirty-four-year-old Craig Harrison, a corporal of horse in the Blues and Royals Regiment in the British army, killed two Taliban machine-gunners in Helmand province in Afghanistan at a range of a staggering 2,475 metres. The weapon he used – an L115A3 Long Range Rifle – would likely not have been powerful enough to reach the Taliban fighters had he shot it at sea level, but because he was at an altitude of over 1,500 metres the thinner air meant the rifle's range was lengthened. At that distance each click on his telescopic sight resulted in a shift of about 25 centimetres as to where the bullet landed. The bullets, when fired, flew for six seconds and dropped about 120 metres on their way. However you look at it, the two kills were incredible feats of skill.[29]

So accurate and deadly have snipers become over the years that they are seen as somehow 'unfair' in war; they are feared and loathed in equal measure. The biographer of Sepp Allerberger, a sniper on the Second World War Eastern Front, described in detail what happened when Russian partisans captured a young German marksman. The youth was dragged into a sawmill and, still alive, had every limb cut off with a buzz saw. His torturers 'tied ligatures around his limbs before cutting them off' so that he didn't bleed out. When Sepp found him, dead, the saw blade was 'still turning, and had reached up to his navel'.[30] In another incident, the Russians found a Nazi marksman with one of their own sniper rifles – its wooden stock full of notches, one nick for each Russian killed: 'They'd cut off his nuts and stuffed them into his mouth. But the worst thing was that they'd rammed his gun up his arse, barrel-first right up to the back sights.'[31]

Knowing the hatred that snipers inspired, I asked this trainer before me, as she sipped her drink, if she had any qualms about teaching such dark arts.

'Yes, sometimes we would sit back and think: "Wow, what are we teaching here?" But we were largely just teaching on a base, all our targets were paper ones. Perhaps not to think of what we were training to do, we even had a stupid sense of fun. We would get T-shirts printed for each course. I remember one of them had very small letters on it that read: "By the time you finish reading this you'll be dead".'

I laughed then, but other T-shirts from Israeli sniper training academies were less amusing. One showed an armed and pregnant Palestinian woman in the crosshairs of a rifle. The caption read: '1 shot 2 kills'. Another had a child carrying a gun in the centre of a target. 'The smaller, the harder', it read.[32]

But she was adamant, like all of those I met who had served with the IDF, that there were very specific rules governing engagement with the enemy here; any soldier who fired without a direct order would face a court martial and go to prison. And she really believed in what she was teaching – she had even married a sniper.

'I got you in my sights,' he used to joke with her.

The next day I headed to the West Bank, home to about 2.5 million Palestinians and 350,000 Jewish settlers.[33] The settlements are considered illegal under most interpretations of international law, but the Israeli government disputes this and gives the settlements its backing.

The Jews who have made their homes here are renowned for being well armed. After all, these communities live under threat from Palestinians who would use deadly force – rightly or wrongly – to shift them off these disputed lands. So I was heading to a military training camp run by Jews that schooled other Jews from around the world in counter-terrorism techniques, hoping to get an insight into some of the armed mentalities that framed this age-old conflict.

The owners of the training camp had called it Calibre 3, and, on arrival, an angry shouting filled the air – the instructors were running a training course. I went around the portable building site huts and peeked through the door. Behind it, a group of pre-teen American Jewish kids were finishing their lesson. The instructor, a towering man, his ripped muscles visible in his neck, was telling them how to stop a terrorist from stabbing you. Beside him a ten-year-old, his mouth full of metal braces, smiled and stabbed his mother with a foam dagger. She laughed and then gave me a fearful look, as this was not my course, and she did not know me. The enemy was everywhere.

I went outside and picked up a brochure for Calibre 3.

'Our classic two hour instructor program,' it read, 'is designed for tourists of any age who would like to get a taste of Israeli methods of shooting and combat.' The images on it were of men with shaved heads. They offered training to security personnel and to wide-eyed Jewish tourists out here on a visit to the old country.

Seeing me, Eitan, a short and trim man in military fatigues, came over. He was the head instructor and told me how Jews from 'all over' came here. Some stayed for as long as thirty days, he said with a thick Hebrew twang, during which they were taught sniper skills, handgun training, rifle handling. The basic aim was to teach them not to shoot 'the good guys', 'only the bad guys'.

'Come,' he said, and we walked around the corner, past a line of high-topped earth mounds covered with camouflage scrub and oil barrels, and out onto a narrow range. On one side were fourteen tourists, all from the US, most in white T-shirts. At the end were paper targets: one of an Israeli soldier, the other of a man with a red keffiyeh, the headdress of the Arabs. Both images showed the men holding semi-automatic rifles, but it was clear which one was the good guy and which was the bad guy.

'From my angle, weapons are designed for killing,' shouted the instructor, a brick wall of a man with Thai boxing tattoos across his arms and up his neck.

'Weapons are not for defending, weapons are for killing. If I want to defend myself, I wear a bulletproof vest, a helmet. But I use this,' he said, lifting his Uzi sub-machine-gun. 'This weapon is for killing.'

'The last time I heard the word "killer",' he barked at the tourists, who were staring wide-eyed at this angry man, 'I heard it with honour.' I could not see, through his mirror Ray-Bans, if he was joking. I assumed not. 'Because that killer killed terrorists,' he said. He was definitely not joking.

The Americans were loving this. An eight-year-old girl in pigtails and a green halter top put up her hand when he asked the group who were the terrorists.

'Arabs?' she said.

He ignored her. 'I am not against Palestinians,' he shouted. 'I am against terrorists. All the terrorists here are Palestinians.'

This was a lecture based on fear, however justified. He called his rifle 'The Devil', and then pulled out an unloaded pistol and pointed it, with one hand, at a man in the front row. The man shifted a little lower in his seat.

'If I shoot this pistol now, who will I kill?'

'Joey!' shouted the girl. Her hand had come down and was now pointing at her brother.

'No!' screamed the man. 'I won't! I won't! I will hit the person next to Joey! See? The pistol kicks to the left when I pull the trigger!'

The person next to Joey looked uncomfortable.

'But if I stand like this,' he shouted, holding the pistol in both hands and legs apart, 'what happens? Who do I shoot?' The muscles on his forearms were throbbing.

'Joey!' shouted the girl again, her pigtails dancing.

'Yes!' shouted the instructor. 'I kill Joey.'

Joey looked upset.

It continued like this for a while. He bellowed about 'neutralising shots in the face from close distance' and how one bullet could kill six people, passing through each person in turn. He screamed that he judged people as terrorists by their actions, not what they looked like. Then he shot the target of the guy in the headdress six times, the bullets clustering in the Arab's forehead.

The atmosphere was febrile. Guns only increased the intensity, the madness. They seemed to make dialogue impossible, and, despite whatever the trainer was screaming, guns here seemed to reduce everything to kill or be killed. It was claustrophobic. Another man, one with sad and intense eyes, came up to me. He was Steve Gar, a South African instructor who had made Israel his home and who was infused with love for his new land. Steve metaphorically carried his rifle in one hand and the Torah in the other. He was one exam away from becoming a rabbi and had, it seemed, spent half his adult life training for a religious life, the other half in the military. He was a man of strong convictions and convincing strength.

He did not like the West Bank being called what it was. 'Why should I define what is Israel in relation to what is west of Jordan?' he asked. 'It's racism.'

He hated the fact that the place where he lived, deep in internationally recognised Palestinian territory, was called a settlement. And his voice lifted in anger when he spoke about how the Palestinians resented the Jews living in their isolated towns. Beyond us stretched a valley of crumbled rock and clumps of scrub, untouched for millennia, the ground here so dry that it could suck up the blood of a thousand armies, and I looked out at it and wondered what was it about this stony land that inspired such passions.

It was a deadly passion, though. He had told me how, as an anti-terror team leader whose job it was to protect Jews living in the West Bank, he had been in at least six serious incidents involving terrorism. I asked him if he had killed, but he refused to tell me.

'Our mission is two things. The first is to protect Jewish life. The second is to protect the Jewish way of life. What they have with the Iron Dome means that one in a million rockets will kill someone here,' he said referring to the air defence system that protects Israel from missiles fired at her territory, 'so I am not worried about Jewish loss of life. But I am worried about them harming the Jewish way of life, for if we bend to them we let them harm our psyche, our psychology. And I want my children to live . . . I cannot blame the terrorists for killing our children, but I can blame them for turning our children into killers.

'We have been running away for thousands of years,' he said, his eyes moist with emotion. 'But when you look at Judaism there is one place where we are safe; it was given to us by God as a promise: Israel.'

But many Jews I spoke to were deeply dismayed about such a gung-ho attitude. In 2008 Israeli prosecutors found that of the 515 violent acts committed by Israelis against Palestinians and Israeli security forces, 502 were by right-wing Jewish settlers in the Occupied Territories.[34]

Perhaps this was Israel's tragedy. The threat to their culture was arguably not from the guns of the Palestinians but from within. It's been said that 'the only democracy in the Middle East has fallen prey to a succession of Right-wing governments, which derive much of their electoral strength from Russian emigres and extremist religious parties'. [35] They have reached a situation where the only way to win any argument is thought to be with a gun.

In 2012, a team of Israeli filmmakers made a documentary called *The Gatekeepers* – it was about the occupation of Gaza and the West Bank. They managed to interview six of the former heads of the Shin Bet, the national intelligence services. Each described the ruthless policies they had once implemented to maintain Israeli dominance in the region and to crush dissent in the Occupied Territories. And most agreed such repressive tactics had been counter-productive. As one said: 'We've become cruel, to ourselves as well, but mainly to the occupied population, using the excuse of the war against terror.'

Democracy can never flourish like this – not at the end of a gun. But Steve was too far down the rabbit hole to see this. Everything here was suffused with an aggressive madness; the gun had become the only way to discuss things.

He summed this up when he told me an anecdote that, in the telling, he had no idea how disturbing it was to hear. He explained how he had placed his own unloaded rifle, and a video camera, in his toddler's bedroom and filmed the child playing beside the gun for two hours, just to see if the child would touch it. The boy did not, said Steve, and then he told me how he had hugged his son and said how proud he was of him for not touching the gun. And, I thought, as Steve showed me how to kill a terrorist – a long-distance shot to the chest and then a close-up *coup de grâce* to the forehead with a pistol – that for as long as men like this had guns in their hardened hands there would always be a problem here in Israel. Just as there would, probably, always be someone else out there wishing to shoot him.

With that thought, I left to get the other side of the story – I wanted to speak to a Palestinian.

The day before, a group of young Palestinian men, the youngest just fourteen years old, had been throwing rocks at the immobile walls of the West Bank in the town of Bethlehem. One of them, frustrated at the mounting deaths in Gaza 50 miles away, had lit a Molotov cocktail and pushed it through a small gap in a metal gateway. The Israeli military had retaliated with gunfire, and another boy, one who had not thrown the home-made bomb, was hit.

I did not know more than this, and so there I was, making my way with a translator to Al Hussain hospital, a medical centre in Bethlehem, to find out more; to try to understand what impact the snipers of Israel had upon the Palestinians.

Arriving at the modest and faded building, we climbed the stained stairs to the fourth floor, where Qusai Ibrahim Abu Basma, a sixteen-year-old student who longed to become a lawyer one day, was recovering from his gunshot wound.

Qusai was handsome and dark-eyed, like so many boys here. His T-shirt had, in capital letters, 'BEGINNING', 'MIDDLE' and 'END' written on it, and, as he shook my hand, I looked down at the dark patch of blood that seeped through his bound leg and wondered if this was the beginning or the end of something for him.

The bullet had passed right through, ripping ligaments and obliterating his shinbone. His father, a silent man who sat spirit-light in the corner, showed me an X-ray that told a different story to the neatness of the white gauze. It spoke of a likely limp and a near escape from an arterial bleed. The boy's mother looked on, shrouded in a hijab. I asked Qusai if he regretted protesting, if it was worth getting shot for.

'We can't do anything,' he said, his face framed in the light of the window. 'They have guns, we just have stones.' He said he would kill an Israeli but he could not get his hands on a gun. 'I was shot because I was the last one there as we left. I was just throwing stones.'

It is hard not to feel something when you meet a child shot for throwing a rock at a wall. And, no matter how much the Israelis speak of bombs and lives lived in terror, you can't equate that with maiming a boy who weighs 40 kilograms at most.

I wondered what would happen if a British or a French soldier had shot this child in Europe. There would likely be a court martial, a prison sentence, all manner of trouble. But this incident was already a day old, and no paper had reported it.

Then he told me he wasn't the only person in his class to get shot, and that the other kid had died. So, after a time, we said goodbye and walked back down the stained corridor. Getting back in the car, we drove to meet the father of the boy. We passed the Paradise Hotel and the Herodian Store gift shop, and, smelling the remnants of something foul in the air – the chemicals used by the Israelis to scatter crowds with their stench – I wondered how far from paradise this place had fallen.

We pulled over at a mural that lined the walls. Some graffiti here showed hijab-clad women toting semi-automatic pistols, Banksy style; or glamorous women with martyrdom in their eyes and AK47s in their hands. But this was different.

'Article 31: That every child has the right to rest and leisure, to engage in play and recreational activities appropriate to the age of the child,' read the text next to five crudely painted footballers. It had been done years before to commemorate Palestinian Child Week.

A short distance from here, eighteen months ago, a boy had been travelling to football practice when a 'disturbance' happened. Perhaps he was throwing rocks, perhaps worse. What we do know is that he was shot with a sniper's bullet.

A human rights group listed the boy as just one of fifty-four minors under the age of eighteen who had been killed between 2012 and mid 2014 in the Occupied Territories. This did not include the deaths of children in Gaza from Israeli air strikes and bombs.[36] The date of his death was 23 January 2013, and his name was Saleh Ahmad Suliman al-'Amarin. He was fifteen.

The house where Saleh lived before a sniper bullet shattered the lives of his family was in the long-established al-'Aza refugee camp. Outside the house were pictures of the boy. The community had embraced his death as that of a martyr, a *shaheed*. 'If you live, live free, or die like the trees standing up,' read an old Arabic proverb across one banner.

My translator rang the buzzer, and, after a time, the boy's father, a thin man with a cracked face, greeted us. He invited us in, leading us to a quiet room at the top of the stairs, lined with baroque, sapphire-hued sofas and memorials to his son.

'He was everything to me,' he said, sitting down on one of the heavy brocaded chairs. 'The promise, the happiness of this house. You cannot imagine. He was the only child I had and now he is dead.' Above him an image of his child looked down.

He stopped and allowed composure to come back. He spoke of the intimate details of Saleh's life. How he excelled at football and was being considered for an academy slot in an Italian club. How his son was so popular that when he died the father had been humbled by how many others had known and loved the boy.

He then said how his only child had been shot in the head, and that it took him four days to die. 'I wish . . . I wish he had a machine-gun,' the father said. 'All of Palestine will take revenge some day.'

Perhaps sensing my twitched response to these apocalyptic words, his tone changed. 'We don't think all Israelis are criminal. If someone is in Tel Aviv, and another Jew here in Bethlehem shoots my son, then that is a different person. I cannot want that Jew in Tel Aviv dead. We are not against Israelis. We are against the occupation here, the Zionists. They take our land, our freedom, our joy.'

I listened to his anguish, and after half an hour we had to go because his grief was washing over him again, and there is not much more you can ask a man who has lost his heart. You can only record.

It was clear that guns, both in this room and in Palestine and Israel, had only wrought the agonies of torment. A thousand sons' lives gone and a thousand fathers' happiness ended: a sorrow repeated endlessly.

Lands where fathers have to leave rifles on their sons' beds to see if they will touch them, lives under the gun's tyranny; where snipers wear T-shirts of pregnant women in the crosshairs; where children are taught to shoot terrorists in the forehead.

These were hard truths. For me, the most militarised nation in the world had so much potential and talent and warmth. Like when, during one air raid warning, a shop owner in one of Tel Aviv's markets beckoned me inside with offers of lemon cake and iced water, apologising for the overhead bombs – as if it were her own fault. Such moments offered small insights into the nature of the Israeli character, and into the eternally complex relationship they have with their country and the violence that besets it.

But it was also the craziest and the saddest nation with the bleakest of all futures that I had ever visited. And I couldn't shake the thought that guns had played the greatest part of all in its endless tragedy.

War and tragedy are ugly twins – co-joined and grotesque. In a way a journalist's life is marked by them – as if drawn towards them like a tourist queuing for a freakshow. And it is never more grotesque than when that show involves children. While I had seen guns aplenty, been shot at, met snipers and victims and seen the ugly face of defence and attack up close, there was one aspect of militaries and guns that upset me more than most. And that was those times when I had met with child soldiers.

One such occasion was the summer of 2012. I had come to West Africa and was travelling on the main road that connects Liberia's capital of Monrovia with the plantation lands of Bomi county. I was there to see, as part of the rehabilitation programmes that the organisation I work for, Action on Armed Violence, carries out in West Africa, what guns had done to children here.

A decade before, this densely tree-lined road was witness to the sort of vicious fighting that so often defines conflict in Africa: young kids with big guns. At its end lived communities of ex-child combatants who, having put down their weapons, were being trained by us in agricultural work – a useful skill for men and women hardened by years of violence.

The potholes dotting the dirt track caused the car to lurch. Over the noise of crunching gears, my driver, Moses, was shouting at me about the best Liberian food to eat for breakfast. He strongly recommended hot pepper soup as a solid start to the day. It was quick and easy to make and left your mouth smarting long enough to stop hunger until lunch. It was a conversation best understood in light of the fact that Moses had, for years, lived with the terrible famine the Liberian civil war had brought to the land.

After two hours of driving, moving under a thick blanketing cloud of ochre-white, the road split, and we took the reddened earthen path up into a plantation. As we left the road, the greenery changed into a broken land of upturned trees, splintered branches and beaten soil. The forest was being tamed. Men stood in the near shadows, swinging machetes at the remaining shoots that had not been knocked down by Chinese-made bulldozers. This jungle was tenacious.

Our car pushed up a hill, a gap in the trees looked down onto a slick of silver – a thin river with five naked boys splashing in the shallows. Then, a glimpse of white ahead, a house deep in the bush. It was the beginning of a village, a loose collection of low-slung huts around shady trees. Corpulent women sat on the narrow porches that fronted their two-room bungalows, slipping food into blackened, bubbling pots. Slim strips of dark wood were being slowly fed into the flames beneath.

As I stepped from the car, a young child broke into deep sobs. The others laughed. 'He's scared of you, white man,' said one. And they laughed again as he dissolved further into a mess of wide-eyed fear and pearly speckled snot.

Settling down, the interviews started slowly. 'Say what the European expects you to say,' I thought. Their answers were short, guarded. Why would they be anything but? But then, as time passed, the conversation relaxed.

Food arrived – bony fish in a brown sauce, served up in a white pot with lilacs etched on the side. Those talking to me – five one-time child soldiers who had grown up to become rubber plantation workers – focused on eating. They dipped their spoons into the mush and ate with fierce concentration, the burned flecks of rice catching the sides of their mouths, more falling onto the ground, where chickens hovered. Behind us, a group of children gathered, one of them holding a slim stick. He began hacking at the forest edge, and the others joined in. Swish, swish, swish; small boys, sharp sticks, milky teeth.

Then, as the food hit their bellies, the four men and one woman who had sat with me in the shade of a sheltering tree began to talk about the war. What war it was – the reason for the fighting – in a sense was unimportant. What was important to me was an attempt to understand what happened when you put guns in the hands of children.

Then the woman among us started talking, and the tone of the meeting changed. Slim, about twenty-five, she showed me a white scar the size of a tangerine on her right shin. She got this wound the day the rebels arrived in her village and started shooting. A bullet had passed through her leg as she lay on the floor of her hut, cowering from their wicked firepower.

Her father had picked her up and carried her into the thorny bush, but the rebels had stopped them before the green had enveloped them, and they had turned on her father, accusing him of fighting for their enemies. The men with the guns had pushed her down and, instead of letting them flee, allowed their own perversions to run wild. They forced her father to carry an impossible weight on his head. But he couldn't lift the bags of weapons they wanted him to porter, so they shot him in front of his bleeding daughter. An AK47 round to the back of his head. Then the rebels took her. She didn't say what happened immediately after that. Many women in that conflict were raped.

But she did explain how, like the others here, she was forced to become a child soldier, because this was what the Liberian rebels were seeking. Those leaders knew that the AK47 was as deadly in the hands of a ten-year-old as it was in the hands of an adult.[37] They knew that children don't eat as much and don't expect to be paid as much, if at all. That children are easier to brainwash and have a less acute sense of danger compared with adults.[38] Such children, armed with guns, are able to commit terrible, terrible acts.

Then she told another story. Some days later these child soldiers captured another man. He was big, and they brought him to her. In front of him, they told her that he looked 'greasy' and that his fat would make a good meal. He pleaded for his life, but they cut him, stripping the skin off him as he screamed and screamed, until one of the young men must have felt a pang of something like humanity and put a bullet in his head. Then the rebels ordered her to carve him up, take out his heart, and make a soup for them.

She set about cutting and boiling, and in the early hours of the morning she had finished the stew. They woke from their sleep and crowded around. Eating the heart of this man, they believed, would make them stronger fighters, it would give them 'jungle magic' – and this meant bullets would pass around them in a firefight. But they feared being poisoned, so the rebels forced her to eat her dish before they ate their fill.

I pushed the fish around my bowl. The others in the group sat in silence, listening to the story. And then they joined in – they too had seen such things.

A pregnant woman, approaching a roadblock, was stopped by two rebel children with AK47s. They argued about the sex of her unborn child. To find out they took a knife to her belly and settled the matter.

A boy once wore the dried breasts and genitals of a woman he had killed with his pistol, adorning himself with these withered amulets so as to protect himself from harm.

Then there were those who drank the blood of other murdered children to boost their powers.

These were stories, or variations of stories, I had heard before. Horror stories know no bounds and I had met child soldiers not only here in Liberia but also in the Philippines, in Colombia, in Somalia and in Mozambique. Just some of the estimated 250,000 child soldiers in the world today, their existence as old as war itself,[39] and they often told the same harrowing tales.

Here, shrouded by Liberia's shading trees, these adults who once had killed as children talked long into the afternoon cool, telling tales of what happens if you make soldiers out of children and give immature hands guns to hold. War descends into a *Lord of the Flies* hell – a charnel house where immature morality is animalistic and where deadly acts are committed without regard.

Their tales got so dark that a doubt flickered in me. I wondered if they had begun to make up things just to shock me. But in the car on the way back Moses told me in a quiet voice that they had only scratched the surface of what had happened those years before in the hidden depths of the bush.

'They were children with guns,' he said. And then he asked if I had enjoyed lunch.

IV. Pleasure

9. THE CIVILIANS

The joy of shooting – Cambodia – shooting AK47s with a beauty queen in Phnom Penh – the American gun owner – meeting a gunslinger in the Midwest, USA – the evolution of gun sports – a journey north to Iceland, a land of many guns and few murders – meeting a sportsman beneath a volcano – understanding the culture of peace through punishment – the rarified world of the gun collector in Birmingham, UK

Guns are fun. I have no doubt about that. When used in the right way and in the right place, they can bring great satisfaction and pleasure. And by 'the right way and in the right place', what I mean is in a controlled and safe manner in a situation that threatens no one.

Of course, with these Faustian caveats a diabolical world of debate is conjured up. People argue endlessly about what constitutes 'control', what is meant by 'safe', and who is threatened. Many believe they *are* in control, they *are* safe. And that they need a gun to protect their liberties against a potentially despotic government.

Civil liberties aside, though, I had to consider guns and their non-lethal use in the pursuit of such things as sport, pleasure and self-defence. After all, in 2007 it was estimated about three-quarters of the 875 million guns in the world were in private hands.

Of course, the numbers of civilian guns varies from nation to nation. From places where every other person has a gun, like Yemen, to lands where there is less than one gun for every 100 people, as in South Korea or Ghana.[1] But overall the numbers of civilian-owned guns are distractingly large. India has about 46 million privately held firearms, China 40 million and Germany 25 million.[2]

As ever, it is the US that takes the prize, with almost one gun per person.[3] Put simply, if you take India, China and Germany out of the equation, the US, with 270 million civilian guns, has more privately owned firearms than the rest of the world put together.[4] No wonder American gun advocates call themselves an army.

But, as with everything related to guns, it's difficult to trust these figures. Millions of small arms worldwide are owned that are never registered with the authorities. Many are registered but may long ago have rusted away or been stolen. And when you have situations like one where a de-activated gun for the film industry in one country, like Germany, might still need to be registered as a working firearm if exported to another in Europe,[5] or where, as in the UK, certain air-pistols are included in the firearm count,[6] figures cited are bound to be an indication only.

One truth, though, remains constant: when guns are used for sports and hobbies, communities evolve around them. And so a group of gun owners and shooters I wanted to consider were those who would gather to shoot without drawing blood; the groups who shoot socially, or competitively; those who keep guns for their peace of mind; and the cliques who expensively and expansively collect guns.

Perhaps above all of these, though, there was one type of shooter who intrigued me: those who do it for the sheer hedonistic hell of it.

Miss Cambodia had, as she said she would, appeared in a wedding dress and was preparing to shoot. The AK47 was handed to her, ready to fire. She raised the stock to her foundation-heavy face and, leaning forward, pulled the trigger. A sharp crack sounded, and at the end of the range, at the edge of a sharply outlined target of a charging warrior, a bullet left its mark. Wounded certainly, lethally possibly. Either way, you knew not to mess with a beauty queen with a semi-automatic.

It was 2001, and I was filming an adventure series across Asia; Cambodia was our first destination. Here the presenter, a journalist with the magazine *FHM*, had eaten deep-fried spiders, had his back 'cupped' by a traditional healer and taken part in a buffalo race. Now his challenge was to beat Miss Cambodia in a shooting competition, and so we were at the Royal Cambodian Armed Forces' 70 Brigade Headquarters – the only firing range in this Indochinese country open to a paying public.[7] It wasn't exactly hard-hitting journalism.

The afternoon was heat-soaked. We had driven away from the genteel lanes of the French colonial capital of Phnom Penh, through a bordered world of rice paddies and fields of dry corn, until the distinctive crack of the Kalashnikov was heard, its sound filling the saturated air and echoing out towards the walls of the jungle.

The other visitors here were white tourists – Australians, Germans, Americans, Brits. It was popular to include a trip to a firing range with a visit to the Khmer Rouge-era S-21 prison. So, in the morning, these travellers had dragged themselves around Pol Pot's interrogation centre – an innocuous three-floored former school, whose balconied walkways and bending palm trees belied its past life. There they stared at the faded photographs of souls killed by genocidal monsters and took pictures of ugly wire torture beds in once fear-filled rooms.

Afterwards they were driven to the Choeung Ek fields on the capital's outskirts. Here Cambodia's intellectual elites, teachers, doctors and journalists had once been taken in their thousands to wet mud flats known then and forever more as the Killing Fields. And the Westerners learnt how Cambodia's leaders had once been killed with an AK47 round in the back of their heads.

After witnessing these horrors, the tourists would then come to this range and shoot guns. Firing thirty rounds from a Kalashnikov cost them $40. For $70 they were able to let rip on an M-60, the US military's machine-gun of choice in Vietnam. There were other packages for more extreme tastes. Lay down $350, and soldiers would drive you 30 kilometres east to a field in the Kampong Speu province.

There a B-40 rocket-propelled grenade launcher would be hoisted onto your shoulder. For another $200 you could fire it at a live cow. I guess if you can buy sex in Cambodia with a twelve-year-old for $10, then someone renting you a weapon to blow up an animal is nothing.

We were just after guns, though, and so in this languid air-force range we had laid down our dollar bills and lined up to take aim. Dogs roamed outside, instructors stretched idly in fly-touched hammocks or hustled at pool. And, in the heat of the afternoon, Miss Cambodia, Samoni, lifted her rifle again.

Behind her a screaming skull from the Airborne Unit insignia stared down. 'Mess with the best, die like the rest,' it read. Vietnam war clichés, I thought, and she fired another round and then another. The paratrooper gun instructor told her to lean into the gun, to expect its recoil. But Samoni had shot before and, as the cameras rolled, her bullets caused the sun-red earth down the range to pop in tiny eruptions, and a smile formed on her rouged lips.

Filming this beauty queen expertly unload a semi-automatic clip into a target was not just a slice of entertainment for television. It allowed us to touch on the terribleness of civil war. You make it engaging so people watch it, as it's hard to film a mountain of skulls and make it popular. Do it like this, and you get 200 million people watching facts about horror and genocide.

The gun was certainly a way to tell that tale – the Khmer Rouge's actions had led to the deaths of nearly 2 million of their own countrymen between 1975 and 1979, many of them with Chinese- and Russian-supplied rifles. These armed fighters, wrapped in their checkered *kramas* and communist ideologies, had spread like a virus from the jungles to the rice fields and then into the suburbs of the boulevard-lined cities, until the entire country had fallen under their control.

These men were not just interested in power. They wanted that thing that you should always fear when men have guns – they wanted to make a new society. So they began to dismantle and destroy what had gone before, in the name of a warped agrarian ideal, and the rule of the bullet was marked by the deaths of countless.

Guns flourished. Certainly you wouldn't have had to come to this military range to fire one. Owning a high-velocity weapon wasn't even illegal until 1998, and by then every third household in Cambodia was armed.[8] But when the Khmer Rouge finally collapsed, the new government realised the revolution's legacy of arms was fuelling criminal violence. So they introduced tougher gun laws and a massive buy-back scheme. Within sixteen years over 180,000 weapons had been destroyed. The local English newspaper, the *Phnom Penh Post*, was to report that guns had declined from being behind 80 per cent of all violence in 1994 to just 30 per cent a decade later.[9]

That's why you had to come to a range to fire one. But shooting with Cambodia's beauty queen was more than just entertainment. There was a peculiar truth present in that moment: a truth of just how quickly guns, even ones touched by genocide, can be used by tourists for fun, how quickly, in a stranger's hands, a gun can shift from something feared to something desired.

Cambodians did not come here to shoot. Only tourists came here to pick up Uzis and Glock pistols and Thompson machine-guns, the remnants of Cambodia's colonial and subverted revolutionary past. The pleasure these guns gave to foreigners was not diminished by what these guns had done.

I understood why. There was a time when I enjoyed guns for the pleasure they brought, and that alone. I used to be the head of a small gun club, a ramshackle affair of a wooden hut and a few tons of sand, where we shot rimfire .22 rounds on bucolic summer days. As a teenager I learned how to fire pistols and target rifles; I was trained to strip down a General-Purpose Machine-Gun in seconds. For a brief part of my youth, I even bought print magazines with names like *Guns & Ammo*. But then I began to travel to war zones and lands touched by the gun's deadly sting, and things changed. Guns lost their allure, and the memory of what they can do to a man took the pleasure from them. And I turned, in the terrified moment of that killing, from enjoying what guns brought to me to looking at the harm they brought to others.

For many, though, their view of guns has never been tainted in such a way. For them the gun is almost a way of life.

'A well regulated Militia, being necessary to the security of a free State, the right of the people to keep and bear Arms, shall not be infringed.'

These twenty-seven words make up the Second Amendment, a constitutional right to bear arms that for many in the US lies at the heart of what it means to be an American. They have had an impact far deeper than was probably ever intended and have led to a country saturated with guns. A 2012 Congressional report found there to be about 310 million firearms there.[10] Of course, not every American owns a gun – about two-thirds of them lie in the hands of just 20 per cent of gun owners.[11] But in the Land of the Free, consumer logic has taken gun ownership to its extreme. Under US law, the National Instant Criminal Background Check System is used to see if someone can buy a gun from a dealer before they can walk out of the shop with it. The figures show that system received almost 157 million applications for guns between 1998 and November 2012.[12]

What those twenty-seven words also created was a situation where the gun is vibrantly visible in American culture. In some states, it's a culture that has evolved into laws that let you carry a gun into a church or onto a college campus,[13] or laws that say it is not a misdemeanour to shoot someone in self-defence if you are drunk.[14] In some places they even permit you to sell a gun to a fourteen-year-old.[15]

It's a right to bear arms that is also vigorously defended. Even modest gun controls – like proposals for background checks – are seen as dire infringements, ones that pave a lacerated route to despotism, because the first thing a tyrant does – so the theory goes – is take away the opposition's guns. It means when a Missourian Republican, in all seriousness, wanted to make it a felony for his fellow lawmakers to propose new gun control laws, he was not laughed out of office.[16] And when the city council of Nelson, Georgia, once tried to pass a law to make it illegal *not* to own a gun, the response of many was support, not shocked bewilderment.[17]

At first, like many Europeans, I did not understand such things. I found it easy to sneer at America's love for the gun. But then I travelled across the broad width of this complicated land and, from diner to boardroom, listened to dozens upon dozens of gun owners. And from them I learned one central reality: that you cannot understand civilian gun ownership in the US without acknowledging the heavy burden of history and violence on the American psyche.

One of the most obvious scars on this psyche can be seen among Americans with Scottish roots. In 1745, for instance, many of their Scottish ancestors were involved in the bloody Jacobite rising, when Charles Edward Stuart set out to regain the British throne for the exiled House of Stuart. Then Stuart, or 'Bonnie Prince Charlie' as he's better known, sailed to Scotland and raised the Jacobite standard in the Scottish Highlands, supported by a gathering of Highland clansmen. He set off on a march south and won victory at Prestonpans near Edinburgh. The Jacobites, in bold spirits, then marched to Carlisle, over the border into England. But their luck changed, and by the time they had reached the then town of Derby, the Jacobite army was faced with the threat of strong resistance from George III's heavily armed English regiments. They retreated north to Inverness and there waged the last battle on Scottish soil, on Culloden moor.

The English army's skilled and devastating firepower won the day and snuffed out Bonnie Prince Charlie's claim on the British crown. It also traumatised generations of Scots. In defeat, many of them left Britain. It was a Celtic exodus that led to the establishment of American towns with names born from sentimentality, like Perth in New York or Aberdeen in Maryland. And these tartan-wrapped communities brought with them the traumatic memories of clearances and English despotic rule.

These memories played an important part in the evolution of American culture, particularly in the southern states; one that has within its heart the central tenets of independence of spirit, the belief in a punitive route to justice and a deep distrust of the meddling of the state, all of which typify your American gun-loving patriot today.

This legacy partly explained, at least to me, why Americans have let forty-four states pass some form of law that lets gun owners carry concealed weapons,[18] and why, after Florida sanctioned the Stand Your Ground law – exonerating citizens who used deadly force when confronted by an assailant – twenty-three other states went ahead and did the same.

It also explained to me why those who have a gun for recreational shooting in the US number some 20 million strong, and why they spend an estimated $9.9 billion a year on their hobby – almost $500 per shooter.[19] Why men and women across America form countless small clubs and societies dedicated to the gun. Hobbyists like those who gather in Arizona's western desert at the Big Sandy Shoot, the largest outdoor shooting meet in the world. They meet twice a year at a stony range a quarter of a mile long and there blaze away with rifle and machine-gun. Over 1,000 targets litter the landscape: some that speed along the length of the range, others that explode in a burst of sparks when hit. And at the end of the weekend, with over 3.5 million rounds fired and their targets lying decimated upon the Arizonan soil, these shooters pack up and go back to their lives.

Then there are those civilians whose passion is for sniping. They sign up to $220-a-day training courses to do this. So popular are these courses that one US company, Robar, based in Phoenix, sold over 20,000 of its $5,000 civilian-adapted RC-50s, a rifle capable of piercing an armoured vehicle from a mile away.[20]

Then there are those who go shooting for ideological reasons: to help America regain a path to redemption. This, at least, is the view of the people at Project Appleseed, folks who dedicate their weekends to the historical tradition of rifle marksmanship, using black powder rifles and historical sharpshooter techniques. They gather in isolated groves and hidden woodlands, and their website shows they are convinced that their guns can renew national confidence, that the grand ideals of America – ethics, discipline, community – can be galvanised through gunfire.

Then there are those communities that gather not just to shoot, but to dress up, cook and enjoy the sheer spectacle of it all. For me this community intrigued the most – because their actions seemed much more about pleasure than power. This was why, with such a choice of communities in the US to meet, I alighted upon the group that both pursued joy through gunfire and also summed up the deep-rooted truths of Americana: the rootin' shootin' cowboy.

Brad L. Meyers prefers to go by the name 'Hipshot'. When you call him this, his face lights up. This Michigan-born southern Californian has been many things in his seventy-odd years – a student, a marine biologist, a carpenter – but his passion is for Single Action Shooting, of which he is president of the national society. This means he enjoys dressing up as a cowboy and blazing away with his pistol.

Hipshot has always loved the Old West; the gunslingers of the OK Corral were the heroes of his childhood. The others on the Single Action Shooting Society (SASS) board – or the Wild Bunch – have similar passions and names. Here in the cowboy shooting community you have to have an alter ego. So there's a Judge Roy Bean, a General US Grant and a Tex – Wyatt Earp and Butch Cassidy have long been claimed. They say their particular form of sport – Cowboy Action Shooting, around since 1982 – is the fastest-growing outdoor shooting sport in the US; they've now got over 97,000 weekend cowboys in eighteen countries worldwide.

'It's a celebration of the cowboy lifestyle. Last year 900 people made it to our End of Trail annual meeting, where we had twelve stage matches taking place,' Hipshot said in a gravelly voice, explaining that 'stages' comprise a series of plates that ring when hit.

Hipshot looked the part, manning the society's stand at a Midwestern gunshow dressed in leather boots, shiny pinstriped trousers, a dapper frock coat and a wild rag of a red bandana around his neck. He stood out starkly among the tactical police gear on sale in the cavernous exhibition, but the required dress for society shooting events is western clothing typical of the time, so there was little else for it. He pointed down to his outfit.

'All of this grew out of a concept of "What would a soldier of fortune have looked like 100 years ago?" Of course, the entire Wild West period was for a very short time – about twenty years, the space between the civil war and when trains started hauling cattle.'

'So why do you do this?' I asked Hipshot, nodding at his garb.

'Oh, we come for so many reasons. Some like the dressing-up; some like the cowboy way, where a man's word and handshake was his bond; some are gun collectors; some are western film buffs; and some are just into the shooting.' They come in their luxury recreational vehicles and in the evenings light up a barbeque, crack open sweating cans of beer and enjoy the silence of the stars. And you believe Hipshot when he says they pose no threat. That there are very few accidents at their meetings. Nobody locks their trailers and trucks, and nothing is ever stolen.

It's a world that sounded a little seductive, as fantasy worlds often do. A world separated from the endlessly sad statistics of some 30,000 annual gun deaths in the US.[21] One even at odds with its own past: Dodge City's first local government law was actually to ban the carrying of firearms – the infamous gunfight at the OK Corral kicked off because Wyatt Earp was trying to enforce a similar law.[22]

But such historical truths did not bother Hipshot. 'It's all about fantasy. But it's a life, too. I've developed lifelong friends, who I'd never have met in any other way. We have a saying: "They come for the gun and stay for the people."'

Camaraderie based around guns is far from new. Gun clubs date back to the Middle Ages; one of the first recorded was the St Sebastianus Shooting Club in Cologne, set up in 1463.[23] But it wasn't until the nineteenth century that the sport went mainstream, with organisations such as the National Rifle Association in the UK leading the way. They held their first competition on Wimbledon Common in July 1860, and so appealing was the skill of marksmanship that Queen Victoria graced that fine day, probably with half an eye towards encouraging the skill of sharpshooting to help keep her Empire intact. She even raised a rifle to her unflinching face and fired the first round.[24] With that queenly shot, the sport of shooting was fixed with a royal seal of approval and its reputation was further boosted nearly four decades later, when shooting was listed as one of the sports in the modern Olympics. It was this community of sport shooters that I was drawn to next.

Today, shooting is a major competitive industry, spawning thousands of national and international sporting events and featuring in every Olympic Games since 1896.[25] Of course, it has undergone significant changes since its early days. Targets were once the shape of humans or animals, but after the Second World War this was phased out for round targets to avoid an association with guns and bloody violence.[26] But back in the 1900 Olympic games in Paris, such sensitivities were far from the organisers' minds. They held a live pigeon-shooting event – the first and only time in Olympic history when animals were killed purposefully.[27] Nearly 300 birds died that day, and the event proved to be such carnage, with the cries of dying birds and rivulets of blood staining the stadium sand, that the stomachs of the audience turned.

If that event was short-lived, the duelling event was even less popular. That competition, held in the 1906 and 1912 games, had the competitors facing dummies, not love rivals, dressed in sombre overcoats. Bull's-eyes were pinned to the mannequins' throats.[28] It was too raw a sight to last.

These quickly ditched sports reflect the wider chaotic history of the early modern Olympics – like when Russia's military shooting team arrived in London for the 1908 games nearly two weeks after it had ended. They were following the Julian calendar, while Britain was on the Gregorian calendar.

Despite these birthing pains, shooting has since emerged as a major sport. In 1896 there were just five events. By the London Olympics of 2012 there were fifteen: nine for men, six for women.[29]

Such shooting matches require intense focus and a skill so exact that marksmen use techniques to lower their pulse to half its normal rate, firing between heartbeats.[30]

It has a history, too, filled with wonderful distractions, like the fact the oldest Olympian was a shooter – the Swede Oscar Swahn. He took part in the 1920 Belgian Olympics at the age of seventy-two, winning a silver medal for the double-shot running deer contest. Or the story of the Hungarian pistol shooter Károly Takács, who, after his shooting hand was badly maimed by a grenade, taught himself to aim with his left hand. He went on to take gold in the 25m rapid fire pistol competition in the 1948 London Games.[31]

However sporting, though, where there is shooting – unless properly contained – there is often death. In 2000, three Colombian gunmen tried to kidnap the former Olympic target shooter Bernardo Tovar and his son of the same name. Father and son were coming back from practising and were armed. The younger Bernardo fired his .22 calibre pistol, killing two of his three attackers and wounding the third.

But is such a thing inevitable, I wondered. Can you have sports shooting without violence? Was the peaceful pastime of amateur cowboys the exception or the rule? Do guns inevitably alter things for the worse? I had encountered so much death on my journey that I was sceptical about whether you could have guns without tragedy. But one place I had read about appeared to prove me wrong.

In the distance was the Esja, a volcanic range whose western parts, formed some 3 million years ago, rose to rain-laden heights. Its summit, 780 metres from the speckled sea, lay hidden from view. A thick grey wedge of cloud had spilled down its sides and covered the sky, and beneath its clouded rim, across the freezing waters of Kollafjörður bay, the sound of gunshot sounded in the whitened air.

I had travelled a short distance from the Icelandic capital of Reykjavík, a city washed in the cold, perpetual light of a northern summer, to attend the final day of the national skeet shooting competition. Thirty-one men were there, and some, dressed in tracksuits with their cold metal shotguns made safe, hanging like broken sticks over their blue and black tabards, waited in quiet clusters. They sat with the patience of men who were used to silence and the passing of slow days.

Others were below, down a small hill. There they stood with focused gazes beside semi-circles of grey concrete markings. The stations from where the clay skeet targets were fired were positioned on each side, and each shooter would, when called, quietly take his place on the concrete roundels. Then they would tense, shout a single call and, following the speeding blur from its trap, shoot the orange discs down in their hurried flight. Across the ground a thousand splinters of clay were strewn – the satisfying remnants of success.

Crack went the rifles and then, if a second target was let loose, a crack would ring out again across the empty bay. Twenty-five targets in each round, five rounds in total – each downed target a point. The shooter who hit the most won; 125 points was the maximum but few achieved this perfect score.

Guðmann Jónasson, a forty-year-old plumber, was up next. He had been shooting for ten years, and the sport defined his life. His fiancée was also a shooter and held the title of national female champion. Jónasson too had won awards. He was willing to sink almost $8,000 a year into this weekend diversion with its English-bought shotguns and 24-gram rounds from Sweden. Today he was not on best form. He had scored 59 points out of a possible 75. The man in the lead was on 67. Telling me this, Guðmann shrugged his tracksuited shoulders and smiled. In a land where a volcano might wipe your home out, there is always that sense of perspective, I guess.

'We have the most shooting grounds per capita in the world. You know Copenhagen?' Guðmann said, rotating an empty Styrofoam coffee cup in his thick hands. 'They have four shooting ranges. And they are a big city.' He paused for effect. 'We have ten.' He looked pleased.

With just over 300,000 people, Icelanders pride themselves on their world firsts. They tell you they have the world's oldest mollusc; they eat the world's healthiest diet; they have the most Nobel Laureates per capita; and the highest per capita consumption of Coca-Cola. This uniqueness was why I had come. Despite the fact that Iceland ranks as one of the highest per capita countries in terms of legal gun ownership, its homicide rate is so low it falls to zero in some years. In 2012, 50,108 people were murdered in Brazil and 14,827 people were killed in the US. One person died violently in Iceland.[32]

At times, the crime statistics office in Reykjavík gets a call from the United Nations Headquarters in Geneva. They are told that their figures must be incorrect; nobody has so few murders. Then they tell the bureaucrat that this is Iceland, and nobody has been murdered for quite a while now. So rare an event is murder that when, in 2013, a fifty-nine-year-old man was shot and killed by Icelandic police, the incident was front-page news for days. It even resulted in a national apology from the country's top policeman. But this is understandable because it was the first time the country's police force had shot and killed anybody, ever.

Yet despite all of this, there are still guns here in Iceland aplenty. About 1 per cent of the Icelandic population belongs to a gun club, and an estimated 90,000 guns exist in this land of ice and fire, with its 320,000 people.[33] I had come to the edges of the Arctic Circle precisely because of this. I wanted to understand how so many guns and so little gun crime could co-exist.

So I asked Guðmann if anybody had been accidentally killed in sports shooting, and he said nobody had been hurt in the gun club since 1867, its foundation. He said an insurance agent had once reviewed his books to calculate the premium a gun club should be paying. But the man had found nothing under the column for gun claims. For amateur dancing things were different.

'Oh, there were pages and pages of accidents for ballet and salsa clubs,' Guðmann said, his eyes twinkling. 'It seems that here in Iceland we are probably better at shooting than we are at dancing.'

Many of the other shooters were indoors; even on a July day like this the weather had sent them inside for thick vegetable soup and layers of butter-lined bread. I walked back up into the modest prefab clubhouse. There a poster showed the classifications of guns: Rifflgreinar cal .22; Skambyssa; Fribyssa .22. The words lingered on your tongue. But what caught my eye was the untended gun rack with four shotguns propped up in it.

I thought of El Salvador, and how guns could never be left like this. Here in Iceland the front door of the club was wide open. There were no guards. But they didn't need tight security here. Everyone knew everyone – Iceland ranks 169 out of 193 countries in the world in population size; a killer has, basically, nowhere to hide. The names on the competitor board paid testimony to this. Jöhannsdóttir, Valdimarsson, Helgason – all Icelandic. It once was the case that, if you emigrated here, you had to adopt an Icelandic name, as laid down by the National Name Council, it was that homogeneous.

I ambled back into the bleached day and down to the range to watch Guðmann. He walked to the centre of the concrete semi-circle and paused, focus visible in his tense back. He called out, and a shot sounded, and the spinning orange clay target exploded in the bloodless air.

Names are a big thing in Iceland.

In the phone book all people are listed alphabetically by their first names. Children address adults by their first names, and adults certainly address everyone else by their first names – even when talking to the president or, as everyone knows him, Ólafur. The phone book then lists the person's last name, their profession and, finally, their home address.

The professional listings, though, are different. They do not have to be backed up with hard evidence that you do what you claim to do. So Iceland has six winners, nine sorcerers, eighteen cowboys, fourteen ghostbusters and two hen-whisperers. In this way, if you pick up a phone book and look in between Jón Heidar Óskarsson and Jón Pálsson, you'll find that Iceland has two Jón Pálmasons. And the one that I wanted to speak to was Jón Pálmason, *skotveiðimaður*, because, while this Jón Pálmason was an electrician by trade, he had a different title here. Here he was 'a man who hunts animals with a gun'.

I had been told that this Jón Pálmason was the best person to speak to if I really wanted an answer to a question that I had. He was happy to talk, and so we met on Bankastræti – the main bar street in Reykjavík – and headed for a coffee. Jón was a handsome man, with silver hair and the complexion of someone who lived in a place with the best diet in the world.

I got straight to the point.

'On the one hand there is Central America,' I said, 'a place where guns are often in the hands of small groups of men who are not afraid to use them. Then you have Iceland, with a tiny homicide rate and yet with one of the highest rates of gun ownership in the world. And my question is – how come?'

He started off explaining that Iceland's low level of violence was partly down to regulation: all automatic and semi-automatic rifles, and most handguns, were banned. Then he explained that acquiring a gun is not an easy process: you need a medical examination, have to attend a gun handling course and then you need to pass a suitability test at the police station.

'But you don't need to have a gun under your pillow here, to protect yourself,' he explained. 'We only need guns here for three reasons. First to hunt, second for sport and third for the very few here who collect guns. No one has them for self-defence. We have no army, no war, and there are so few people here that I think it leads to a sense of peace.'

There was, he said, a tradition in Iceland of pre-empting crime issues before they arise, or stopping issues before they can get worse. Right now, police were cracking down on Hell's Angels gangs, while members of the Icelandic parliament, the Althingi, were considering laws to increase police powers in the investigation of their country's small biker networks.

It seemed that dialogue formed the basis of order here. A few hundred metres from where we were sitting there was a monument.

They called it 'the black cone' – a split rock that served as a tribute to civil disobedience.

'When the government violates the rights of the people then insurrection is, for the people, the most sacred of rights and the most indispensable of duties,' a plaque beside it read, quoting the eighteenth-century philosopher Gilbert du Motier, better known as the revolutionary Marquis de Lafayette. This was not far from the American fear of a despotic government taking control. But the US response was to buy guns in anticipation of Armageddon. Here in Iceland they insisted on democratic process and tolerance.

Much of this was food for thought. But I still wanted to get to the root of how such a tolerance and measured response could have evolved. So, after meeting Jón, I drove out of the capital and into Iceland's riverine world of igneous rock and streaming rain.

The man behind the desk at the car-hire company had joked it had only rained twice this year in Iceland. 'The first for twenty-five days, and now for seventy-five days.' I laughed then, but the torrent was now so bad that my windscreen momentarily flooded, and I felt like I was drowning. On either side stretched an undulating world of soaked moss and ancient tholeiite boulders, and my car sluiced through along the road-river.

Forty kilometres out, and, unexpectedly, it rose before me: a hard wall of rock, overlooking a landscape of arterial brooks and eddying rivers. It was the Þingvellir – the ancient Icelandic parliament. Around it dwarf birch trees and moss campion sprouted from craggy rises and purple sandpipers, their backs a spread of diamond rain drops, flew into the whiteness at the sound of my approaching car.

Here, centuries before, the Icelandic government had come to pitch their summer tents at their annual meeting. It was a place steeped in ancient rulings. In the rocks, you'd catch glimpses of cut-deep blackness and wonder what secrets lay in those shadowed ravines. For me, it offered a small insight into why Icelanders seemed able so respectfully to accommodate the gun laws handed down to them by the state. Hard punishment had once ruled those ancient days and blood had flowed here.

'In Olden times,' a metal panel read, 'drowning was widely used as a method of execution. People were drowned in marshes, in fresh water and in the sea. In Iceland, provision was made in law for execution by drowning from 1281.'

Names dotted this rising landscape, loaded with the public threat of what would happen if you crossed the line. In the Stekkjargiá Gorge was the Scaffold Cliff, an islet on the Öxara River formed part of the Execution Block Spit. Another place, Brennugjá, held the Stake Gorge, where sorcerers died in the blistering heat of an auto-da-fé. And then there was the Whipping Islet.

These were hard punishments for hard times. Such public displays of state violence left their mark so deeply on the collective conscious-ness that by the nineteenth century Icelandic justice had less and less need to instil fear – a social contract seemed to have been born. Things become progressively less punitive. By 1928 the death penalty was done away with. And today a life sentence with eligibility for parole after sixteen years is the most extreme punishment dispensed.

State punishment, something so brutal it once marked its name on the landscape, became infused with humanism. And this shift has impacted the whole of society. Today Icelandic police officers do not routinely carry any guns.[34] In the first decade of the Icelandic drug courts, about 90 per cent of the cases were settled with just a fine.[35]

It was as if justice had proven itself here, where it was understood that heavy-handed violence was never going to work things out – by either state or people – where executing someone was never going to stop murders; and where state-sanctioned use of guns inflamed, rather than reduced, their threat.

None of this, of course, explains entirely why Iceland has so many guns but so little gun violence. I'm sure there is something about the hypnotic landscape or the sheer isolation that touches and pacifies people's lives here. But tolerance, liberal state punishment, the right to public dissent, gun control – they must all play a part in a world where civilians can own guns without bloodshed and sorrow.

There was one final kind of civilian gun owner worth looking at: the ardent collector of guns, someone who collects not for self-defence but for the love of history and design. He belongs to a breed of men (and it seems only to be men) who can afford to spend $89,000 on a pair of duelling guns. He lives in a luxurious apartment in Paris or London, has a grand country home and a hidden Swiss bank account. He collects in secret, in part because it is a little gauche to display one's wealth to the masses – that's the best way to pique the interest of the tax man – and partly because telling someone you have a million dollars' worth of firearms in your living room is a sure way to get burgled.

I had gone to the British Shooting Show, a few miles out of Birmingham, to catch a glimpse of this world. It was, in many ways, a gun-lite show. There were, of course, guns aplenty, but it had none of the paranoia of an American show – none of the tactical police garb, no black semi-automatics, no hyper-militarism. Instead it was all hunting dogs and tweed, wellies and country sports, ruddy-faced pastimes that spoke of a bygone age and honey still for tea.

In one corner were rows dedicated to the sale of antique guns, and there one name stood out from the others: Bonhams, one of the world's oldest auction houses. The stand was manned by Patrick Hawes, the head of department of its modern sporting guns section, but he was too busy evaluating guns to talk to me.

Robin Hawes, though, was able to talk. He was Patrick's father and had the look of a charming Georgian parson, or an officer in the Crimean War. He had a trim figure, a strong, creased face and, beneath it all, a roguish glint. This raconteur was one of those special sort of old Etonians who never really grows up and who is universally liked. To me, he summed up the spine of the shooting classes and the heart of Middle England's lower upper classes – a very specific niche which he personified perfectly. He was also clearly in love with the allure of guns and history.

'It's a fascinating subject,' he said, 'essentially a lot of "boy's toys".'

Such things ran in his blood. His father had a pair of renowned Purdeys, the best of the best British shotguns, and on his eleventh birthday Robin had shot a squirrel with one of them. He remembered the recoil being so strong that he fell flat on his back. From then on his life was touched by gunmetal. He spent time in the Grenadier Guards, then at the London Stock Exchange, then at a wine merchants, before ending back with his one true love – antique gun trading. He worked at Holland and Holland, William Evans and Bisley, the cream of Britain's shooting world, and clearly, with his charm and his Eton-Army officer provenance, he was the type of chap you'd like to have with you on a shoot, if that's what you are into. Today he collects for both pleasure and investment. And they make good returns: valuations for hand-crafted English shotguns climb at almost 5 per cent a year, a safe equity market bet by traders' reckoning.[36] So I asked him what guns people liked in this world, what they looked for.

'Some people desire wheellocks, flintlocks, percussion guns. But the provenance of the weapon is really the major factor. If it has belonged to a king or a queen there would be much interest. And that's what makes the antique arms and armour trade worth many millions.'

This was no surprise. I'd already read that Hitler's golden 7.65mm Walther PP sold for $114,000 at a 1987 auction to an anonymous bidder. Today it would be worth many times that. I also knew that Teddy Roosevelt's double-barrelled shotgun once went for $862,500. That gun certainly had provenance. The moment Roosevelt left office in March 1909 he had set off on a year-long African hunting expedition, with an entourage of 250 porters and guides. Thirteen months later, his party had killed and trapped about 11,400 animals, including six rare white rhinos, eleven elephants and seventeen lions; many of the kills with this gun.[37] Teddy called it 'the most beautiful gun I have ever seen', and its sale generated massive press interest, not least because it came with some interesting cleaning cloths: torn slivers of Teddy's pyjamas. Never underestimate the allure of the American hunting legend.

But perhaps the most expensive gun out there is the tiny 6 inch derringer that ended the life of Abraham Lincoln. His killer, John Wilkes Booth, paid about $25 for it in the mid 1800s, but it now resides at the Ford Theatre, where it changed the course of history. Its value is referred to as 'priceless'.

In front of us Bonhams had set up its stall – an empty square of glass-topped tables. Robin beckoned me forward with a nod and began to show off some of the guns on display: a pair of ugly Danish breech-loading percussion guns; some German wheellocks and flint-locks. The displays here were not the only draw. Others came to sell. They sidled up surreptitiously, reached into gun bags and pulled out ancient shotguns for an evaluation. Robin was polite, but his eye was quick to see quality, and there was not much here.

I asked him if he was expecting a gem to turn up today, and he waggled his head in indecision.

'Not really, but you never know. There are certainly fewer earrings, tattoos and pot bellies than a few years ago.'

With those few words Robin summed up to me a nuanced world of class and desire, where the British landed gentry sought heritage and quality not only in their horses and marriages, but also in their guns. Provenance was everything.

It was also a thought that leads me next towards another class-steeped world: that of the hunter, a world where money, breeding and guns seem entirely to define a lifestyle.

10. THE HUNTERS

Somaliland – the ancient art of hunting revealed in a cave – an outdoor show in Germany's Bavaria – meeting hunters, sellers and killer dogs – the trophy-lined showroom of one of the oldest gun companies in the world in London – on safari in South Africa's Eastern Cape – a death and remorse

This stony, arid land had once teemed with wildlife: spotted hyenas and hyraxes, leopards and Barbary lions, cheetahs, oryx and dik-diks. But then the guns of civil war had sounded in the closing decades of the twentieth century, and these animals were hunted and hunted again to feed the Somali troops that fought in these ochre desert lands. Now the plains were silent.

The war in the late 1980s had been a brutal one. Somaliland, seeking to liberate its people from the brutal governance of Siad Barre, Somalia's vicious dictator, had begun to establish itself as an autonomous republic in the far north of the Horn of Africa. Barre had responded with hard force; about 350,000 people died from the violence and the famines that followed. Countless animals died too.[1]

So today you see mainly bush pigs and camels roaming the desert sands, and the only reason the pigs survived was because Somalia is a Muslim country, and the pigs were considered *haram*.

We weren't interested, though, in the hunted of the present. The hunters of the past drew us here. We were speeding across this thirsty earth to see some of the oldest rock art in northern Africa and, as the tourism minister of Somaliland, Mohammed Hussein Said, was with me, it was not proving to be your conventional tourist trip. First, there were no other tourists here, and had not been any for months. There were also six men with AK47s guarding us.

There was an underlying fear that Islamic groups would target Westerners here, wanting to grab a hostage or make a theological point in blood. A few days before, working on a BBC documentary, the crew and I had met a group of suspected terrorists in the main prison in Hargeisa, Somaliland's capital. They had been paraded in loose manacles before us, and all had refused to return our gaze. They were said to be Al-Shabab gunmen, responsible, a few months before, for the shooting and killing of Richard and Enid Eyeington, a British aid couple in their sixties. The Eyeingtons were good people who, according to their friend, the late film director Richard Attenborough, were 'an inspirational couple, selfless and courageous'.[2] If affiliates of Al Qaeda would do that to a gentle couple, what would they do to a bunch of journalists?

The tourist minister was taking no chances; our escorts went armed.

After an hour's drive, the dust blinding and coating, we saw a smooth granite escarpment rising from the north-east. Up there, after a scramble across ancient boulders, the minister showed us caves that contained some of the earliest paintings known to man, some dating back 11,000 years. Ten caves contained scenes of striking antediluvian beauty: animals captured in elaborate ceremonial robes, necks embellished with white plastron, hunting dogs and wild animals crouched beside them and hunters stood beneath them all, their arms outstretched.

'Under here,' said the minister, 'you can see the person. That means they were praying. They say these pictures are the best they've seen.'

He was right, this intense and refreshingly unpolitical man: the paintings were impressive, and not just to archaeologists. Here was art speaking to us of ancient hunting skills and the deities that watched over the kill. In these dry lands, and to the south-west in modern-day Tanzania and Kenya, man had first learned to hunt for his meat, long before he had learned to paint for his gods. Even today Bushmen hunters taint their arrows with the poisonous exudate of the Diamphidia beetle – so powerful a mere scratch can kill a kudu or a man.

Africa's history is, in this way, not revealed through soaring cathedrals or brooding sky-framed fortresses. Rather, dry and secluded caves reveal the continent's historical past in bones laced with stone-tool cut marks. Other caves revealed how the bow and arrow or spear were first used here tens of thousands of years ago. And in those caves, where there were remnants of painted animal gods, you saw man's love of meat, hunted and cooked.

What these vermillion and citrine paintings really showed me was just how deep-set hunting is on our collective psyche; how it has been intrinsic to the evolution of societies, cultures and religions; and how man, underneath our civilized carapace, is a hunter to his core.

The way religions around the world approach the issue of hunting reveals how societies see the world. Many Buddhists and Jains believe that every life is sacred, so they do not hunt. Jews and Muslims do hunt, but generally only for food, not for sport. And they are both forbidden to hunt other trained hunting animals, such as birds of prey, possibly echoing doctrinal views that food needs to be kosher or halal before it is eaten. Christians can hunt, the only exception being Catholic priests, who are not allowed to – a position that sounds very similar to the Vatican's attitude to sex.[3] And Hindus are positively encouraged to hunt, their scriptures describing it as a sport of the kingly. Even the god Shiva is called Mrigavyadha, 'the deer hunter'.

As with religion, hunting reflects people's own views of what is right and wrong. Very few argue that societies which need to hunt for their basic survival are morally wrong for doing so.[4] But many have a problem with people who go hunting for leisure or pleasure. The social-media storms of bile that ensued after an American huntress was photographed posing with a downed lion testify to that.[5]

I take the middle ground. I cannot see why you need to hunt a leopard that is not posing a direct threat to your community. But I have little problem with hunting bountiful, normal game, provided the meat is for the pot. I certainly can't understand those who gladly eat hamburgers or lamb chops but object to those who hunt properly husbanded animals.

Having seen the ugly, mechanised slaughter of modern abattoirs [slaughterhouses] up close, I see no moral difference between meat butchered there and meat that's been hunted.[6] Both, admittedly, are not necessary for survival, but vegetarianism is one virtue I've yet to embrace.

Of course, I'm not alone in my love of meat. Fifty years ago global meat consumption was 70 million tonnes. By 2007 it was 268 million tonnes, a rise of almost 300 per cent. In 1961 we ate about 22 kilograms of meat per person; by 2007 we were eating 40 kilograms.[7] This rise in meat consumption occurred as the world saw a huge shift in urbanisation. Today over 50 per cent of the world's population live in towns and cities,[8] and given that some cities have only 8 per cent of the bird species and 25 per cent of the plant species to comparable undeveloped land,[9] it's clear that billions of people are consuming meat without really living in sight of the natural world from where it comes. Despite our deep ancestral links to a shared hunting past, the death of animals for food in many parts of the world is witnessed only by those who work in slaughterhouses or by those who hunt.

I count myself among those who know neither world. I live in a large city and do not have a garden. I only really see nature when I escape the chokehold of an urban ringway.

This is not to say I had never hunted before. I had sought to catch sharks with lassos in Papua New Guinea. I had been on crocodile expeditions armed with spears. And I had butchered ewes and pigs. I had never, though, shot an animal for the sport as well as for the meat. The other worlds of the gun had shown their faces to me long before this book was conceived. The world of the rifle hunter, though, was totally new to me. I had much to learn.

There is the battue, or the beating of sticks, where you drive animals into a gun's range. You can go calling – the art of mimicking animal noises to lure them to you. There is blind hunting, which involves waiting for animals from a concealed hide. There is stalking, the practice of quietly searching for your prey. Persistence hunting is done by running your prey to exhaustion. And then there is netting, trapping, spotlighting and glassing.

These things I did not know, but I was trying to learn them as I battled with an immense German breakfast of *Aufschnitt und Käse*, in a functional hotel on the outskirts of the Bavarian town of Nuremberg. In an hour the IWA Outdoor Classics – an international show for hunting guns and a wide range of shooting sports – was opening. Given it was filled with huntsmen, it felt appropriate to fill myself with heavy meat and to learn the secret language of poachers and trackers while so doing.

About 40,000 hunters from over 100 countries had travelled to this town in southern-central Germany, and it was, simply, the best place for a novice like me to be introduced to the worlds of chasers and deerstalkers.

Finishing my plate of sliced sausage, I headed to the subway. Nuremberg's metro was teutonically punctual and shifted smoothly through this quiet town. The air was brisk and cold, the sky a blank white. Outside, large functional and faceless office buildings sped past. From time to time men in smart suits pulling petite roller briefcases joined the train; they were on the way to the shooting show, too. Gun sellers.

A Frenchman boarded, wearing a light-brown moleskin jacket and fox tie, speaking loudly into his phone. Behind me two Englishmen with broad faces and even broader Cornish accents talked about profits. The train slid to a stop.

As we stepped out, the sprawling convention centre lay on all sides, up a covered slope. It was a broad cathedral to commerce, struck in minimalist white and glass and steel. Thousands of people were striding with intent, and a pianist played the Roberta Flack classic 'Killing Me Softly' inside. Armed with a press-tag lanyard, I walked in.

The first stall was not discreet, despite its attempts to be so. J. P. Sauer & Sohn. Established in 1751, it was clearly a gun manu-facturer that wanted you to know where you stood in the order of things. A large and largely empty VIP section was cordoned off behind a line of glass cabinets, in which stood the manufacturers' history in gunmetal – shotguns from 1899, 1894 and 1885, the earliest a Kal.12 Perkussion-Doppelflinte from 1835. It had an exqui-site hunting scene engraved on its stock, and these detailed intricacies spoke of privilege and clear class lines. Yet, for all their implied elitism and advertised good taste, J. P. Sauer & Sohn were still willing to make expensive kitsch. There stood a Steampunk rifle, decorated with fanciful flourishes in an attempt to capture a sense of apocalyptic industrialism, sold to a man whom the Sauer rep disdainfully called 'an Arab buyer' for $154,000. Beside it lay a bespoke 'Genghis Khan' hunting rifle, engraved with mystical symbols from the East, its stock a maze of Chinese exotica. It was mounted on a dais beside a Moorish helmet laid out upon a bed of desert sand. It had the price tag of $119,000 and looked like a Disney nightmare.

It was clear this was a show where money talked, where it could buy you a lesson in prescribed taste or cause the self-same arbiters of etiquette to look the other way. So, if you took your Genghis Khan shooter from its rifle case and signed up to an over-priced safari the people here would, you imagine, follow the lead of one of the advertising flyers and insist you were 'hunting in elegance'.

I carried on through the high, vaulted space. There were huge stalls, each themed, each drawing your eye with carefully considered marketing allure. Despite the early hour, people were sipping white wine at a Scandinavian bar made of bleached wood, pelts and furs. Opposite them was a tent for the British shotgun manufacturer John Rigby. It was a safari fantasy in zebra skins and wicker chairs. Even its whisky spoke of a rugged past: Monkey Shoulder – a nickname given to the lopsided look that maltmen once got after endlessly turning the barley by hand.

Then an enormous wolfdog, a 30-kilogram monster with amber eyes and a high-set tail, walked past. A beast that came out of a 1950s programme that sought to merge German sheepdogs and Carpathian wolves, it had been bred as an attack dog for the Czechoslovak special forces. It turned and stuck its muzzle into my crotch. I gently pushed it away, and it followed me until I darted behind a screen to come face to face with a woman in a white shirt and a high-waisted black skirt tapping meticulously on a long barrel fixed firmly in a vice. Her movements were steady and sure, and I stood momentarily enamored of the delicate skill of her art. She was marking an elaborate design into the metal upper housing of a shotgun. Small circles showed upon metal and were expanding under her guiding hands into a subtle rococo flourish.

Her name was Lieben. She was in her mid thirties and had been an engraver for thirteen years now, her role one of the chief engravers for the Belgium arm of the shotgun manufacturer Browning. She was based in Liège, the centuries-old home to gunsmiths, and worked in a unit custom making about 100 bespoke rifles a year. She could tap like this for hours, she said, taking her craft seriously and slowly. One day she might like to do knives or jewellery, but for the moment she was content with beautifying firearms. Lieben found this gunmetal work meditative, her attention to detail part of a tradition that spanned back to medieval Walloon artisans.

'Let Our Craftsmen Create The Gun Of Your Dreams,' said a Browning advert on her right. Twelve workers could take up to a year to produce bespoke guns like these – the engraving alone on a rifle takes 500 hours.

The head of sales for the custom shop walked up. His name was Lionel Neuville, a youngish man with a filled-out face and a high forehead. He was clearly in love with the artistry of the weapons here. When he was a child his mother once took him around the grand museums of Europe. Now, though, she's mystified at his hunting lifestyle, hating the idea her son sells such instruments of death. But it seemed to me it was more the allure of beauty in the gun, not blood, that had inspired Neuville's enthusiasm.

'I'm not really here to sell guns, more to show the quality, the feeling of the gun – what is possible,' he said. 'Our customers are mainly self-made men and are really into stunning things: guns, cars. They want a "wow" item to show to their friends, and we are that: the Aston Martin of the gun.'

His usual client is a rich man in his mid fifties who has been hunting for a while. He invites them over to Belgium to view the factory; they are chauffeured from the airport and given a $550 luxurious French meal to seal the deal. He pats his stomach, and you can see he's sealed a few of them. One thing that he does not do, though, is discuss the price with the client. That would be gauche.

Of course not all customers want the same thing. He can't sell gold-plated pistols in France – 'It's a bit "bling-bling"' – but the Germans or Americans buy such things happily.

'The French and the English may not like each other, but they are similar in that they know what they like – even down to the colour of the wood,' he said. 'The French want a yellowish wood, the British a dark red.' Such bespoke taste does not come cheap. The double-barrelled shotguns he sells run into the upper tens of thousands of dollars. But for many of the buyers, price is never the issue.

'The Russians have the money, but they want to buy a piece of history,' he said. 'They like to buy the Side-by-Side because this was what the Tsar had.' A shotgun made for Tsar Nicholas II recently sold in auction for a record $287,500, so it was clear the Russians were prepared to dig deep for the right image.[10]

'They want people to see their guns, but they will be discreet about it. It's not "take a look at my new gun", but they leave it purposefully at the entrance to the house. It starts a new conversation.'

Of course, there is a long, deep history of European hunting with guns. The Lithuanian, Finnish, Czech and Polish national hunting associations all recently celebrated ninety years since their founding, while the granddaddy of the Union of Hunters and Anglers in Bulgaria had its 115th birthday in 2013.[11] By and large, European hunting is incredibly popular. Today there are about 7 million European hunters.[12] Finland, for instance, has the third-highest rate of firearm ownership in the world, and over half of its firearm permits are for hunting.[13] There are an estimated 1.3 million hunters in France, and 980,000 in Spain, while the island state of Malta has the highest hunter density of anywhere in Europe – possibly the world – with fifty hunters per square kilometre.[14]

It is even more extreme, as many things tend to be, in the US. There, pro-hunting groups claim almost 14 million Americans hunt every year (some say as many as 43 million Americans hunt, but that seems to be overstating things);[15] 58 per cent of all those who carry guns reportedly do so for hunting;[16] and there are over 10,000 clubs and organisations across the US dedicated just to hunting, such as the Safari Club International, the National Wild Turkey Federation and Ducks Unlimited.[17]

Not surprisingly, it's big business and always, in a sense, has been. In the early days of the American frontier the hide of a deer was worth a dollar – which is how the term 'buck' for a one-dollar bill came about.[18] Today, American hunters are said to spend 38.3 billion bucks on their passion, more than – the hunting lobby claims at least[19] – the revenue of Google.[20] Hunting supports an estimated 680,000 jobs: the $26.4 billion in salaries and wages being larger than the entire economy of Vermont.[21] And it's reportedly growing: between 2006 and 2011, the number of hunters was said to have increased by 9 per cent.[22]

Of course, with this much money and enjoyment at stake, the US has a very strong political voice that shouts loud about the benefits of hunting and the right to bear arms. They argue it is safe and humane, environmentally sound and economically beneficial. Some disagree, clearly. According to the International Hunter Education Association's own historical data, over 1,000 people in the US and Canada are accidentally shot by hunters a year, with about eighty of those accidents being fatalities. Hardly safe.[23]

Amid Nuremberg's artisanal beauty, though, you could hardly envisage danger and death. Here was luxury and life. The sections spun off into different gun genres. Some focused on selling decoys and targets of ducks and geese; others sold high-quality ear protection; some just made their living by selling gun-care oil. There were multi-pull clay pigeon systems for training, or shooting-range ventilation manufacturers (yours for $33,000). A Scottish-based gunbox manufacturer from New Zealand would make you an elaborate storage system in walnut for $8,800. There were optic sights and gutting knives, thick stalking boots and wrap-around sunglasses. Everywhere was a microcosm of economic supply and demand – accessories for weapon and hunter alike.

Robin Deas, an old-school Brit, summed this up. He showed me how he had built a flourishing business by focusing on a very specific aspect of gun-hunting culture: namely, feet. At seventy-three, he runs the House of Cheviot and sells knee-high, luxury stalking socks to the hunting classes: merino wool reared in Australia, spun in Italy and knitted in Hawick on the Scottish borders. And he told me, in a crisp English accent, that these socks, in the colour of cinnamon and moss and bilberry, sell for as much as $445 a pair.

'To presidents and kings, sultans and queens', he said with a smile. I was sure most of them wouldn't want the world to know their socks cost close to their subjects' living weekly wage, but it was testimony to the rich micro-climate of the hunting economy.[24]

I wandered on: into a world of skull mounts designed for bleached trophies from East Africa; enormous bronze sculptures of stalking leopards; Slovakian hunting furniture built in an explosion of jutting wood and wrought iron; and Italian stands selling delicate silver trinkets of pheasant and fox. It was a marathon just to wander the endless aisles.

Slowly, a feeling of claustrophobia began to grip. The entire place was so focused on hard selling and slick marketing that the romance and open space promised by the hunt diminished and died. Here was a good place to see merchandise: luxury, bespoke and of the highest quality. It was also a good place to be an anthropologist studying Europe's landed elites. But it was not a good place to understand the motivation of the rifle hunter.

Loaded down with brochures and name cards, I walked wearily back to the metro. But my stalking had paid off; I had caught what I set out to get: an introduction.

The office was in a run-down street in a run-down part of my home city. I walked past a fly-poster-daubed corner shop, continued opposite a washed-out council estate, then, skirting a line of under-the-arches businesses offering cheap car repairs, came to a bleak south London side road.

It was still a far cry from the open plains of Africa or the medieval charms of Bavaria, but I had travelled a short distance from my home, heading south of the river, to talk to Marc Newton. I had first met Marc at his African fantasy pavilion at the IWA show, and as he was the managing director of John Rigby & Co., one of the oldest gun companies in the world, I had asked to see him again.

He had agreed, knowing that I wanted to talk about the culture of the huntsman and he was well placed to have such a conversation.

He oversaw a venerable company that had, over the years, become renowned among big-game hunters for its powerful guns, particularly among a certain type of larger-than-life American hunter. It was the gunsmith of choice for bold whisky-touched men and for crown-touched kings. They had made rifles under Royal Warrant for the last three King Georges and for Edward VII.

Rigby were in the middle of a refit. As I rang the doorbell, the buzz of drilling and the murmur of polishing buffers sounded through the door. Inside, the décor was evolving; it promised to be in sharp contrast to the industrial grime outside. Here you were spirited away to a privileged world of deep-shine leather and aged spirits drunk from fine crystal.

'Everything is here for a reason. We went for a colonial, East African feel,' said Marc, pointing at the heads of impalas hanging from the wall.

Marc was an outspoken and yet disarmingly engaging man, and surprisingly young for the head of such a historical gun company. But then again, his father and his grandfather had both been game-keepers, and he had been raised on the virtues of hunting.

He spoke plainly. 'Let's call a spade a spade. There is bullshit about guns in all their aspects. Something about guns seems to empower people, giving them a right to lecture others. Gun is just one word, but it says a million words. It's one of the most emotive words in the English language,' he said, settling into a leather high-backed chair.

I asked him what guns meant to him.

'Hunters love guns; we have a deep passion for fine-quality items. We enjoy hunting because it's so different from the society we live in, where we are trapped in front of computers. A fine hunting rifle is your ticket to transforming your dull life into those scenes you see in these black and white photos – back to a time of adventurers. When someone buys a Rigby they buy into that image, a key to that lifestyle. On a Friday night they can transform themselves into Denys Finch Hatton.'

Finch Hatton, an old Etonian and Oxford-educated aristocrat, was an interesting example to use. He was a big-game hunter, who, when on safari with the Prince of Wales, later Edward VIII, was asked to creep up on a rhino and stick the king's head – taking the form of postage stamps – on its bottom. He did so, one for each buttock. When Finch Hatton died in a plane crash in 1931 his brother had a quote from Coleridge inscribed above his grave: 'He prayeth well, who loveth well both man and bird and beast.' I was not sure that most bankers picking up a rifle on a weekend had such noble sentiments or romantic sense of style, but I got Marc's point.

Here Marc was selling a lifestyle as much as a weapon – the life of the big-game hunter in South Africa, a whiff of Hemingway in metal and wood. It was, at the very least, a good business tactic – because even though they sold only about 250 rifles a year, some went for $100,000 and more. The most popular of their line was the 1911-designed .416 Rigby, a $20,000 rifle that could take down an elephant. All of this epitomised, he told me, a certain spirit.

'It's a real roll-up-sleeves culture,' he said. 'What I find fascinating is that people look at me – a young man going out stalking, shooting, butchering – as barbaric and macho. Yet the same people see a *National Geographic* video, and all of a sudden that is "cultural". But to that I say: "What about my culture?"'

Of course, this being England, I could not help but think that this culture was one notable for its privilege. Even this room, with its bespoke furniture and leather-bound books, spoke of inherited wealth or city bonuses. But it was not the first time that people I'd met in Britain had been defensive about their right to hunt. People were quick to claim they and their sport were misunderstood, even persecuted.

One bluff Yorkshire huntsman had sounded indignant on the phone when I suggested that hunting cultures in Britain were exclusive, even though his own website showed only photographs of middle-aged white men dressed in matching orange-brown tweed, sitting in leather chairs enjoying a post-shoot drink. He told me, in a hectoring way, that the people he employed to work on a grouse shoot were from all walks of life. Then he reprimanded me for describing shooting grouse as hunting.

Many, though, have a perception of hunting as the sport of the British gentry, one as much about class as it is about animal welfare – a measure of status. It is something underlined by things like press interviews where Princess Michael of Kent once claimed she was experiencing economic austerity, too: 'I sew better than any nanny we've ever had . . . And my father had a farm in Africa. Have you ever taken the insides out of a stag?'[25]

It's not surprising, then, that left-wing polemicists saw class lines when the British Conservative-led government kept the cost of gun licences at $74, resulting in a government subsidy of some $25 million a year, just as they saw privilege at work when the subsidy for grouse moors was increased from $45 per hectare to $83. The left-wing critic George Monbiot wrote at the time: 'So back we go to the hazy days of Edwardian England: a society dominated by rentiers, in which the city centres are set aside for those with tremendous wealth and the countryside is reserved for their bloodsports . . . our money is used to subsidise grouse and shotguns.'[26]

But, for me, such issues spoke more about the hoary British class system than guns. So I steered the conversation with Marc back to hunting and asked about the controversies around big game – the horror many feel when someone is photographed kneeling next to a felled leopard or cheetah.

'Anyone who can shoot a beautiful animal like this one,' he said, pointing to the skin of a lion on the wall behind, 'anyone who can do that without having a pang of guilt – well . . . I feel guilty.

'But,' he said, and I was expecting the 'but', 'there is a use to the flesh. Within half an hour an elephant will be chopped up. We shot a hippo two years ago, and out of the bush people just appeared and chopped it up into pieces, and the meat went back to the local people.'

I did not buy this argument at all. Hunting for a deer, sure. Hunting for a lion that has killed someone, fine. But killing a lion just because you wanted to: I couldn't understand it. I was pretty sure most didn't kill them just to eat their steak.

A few months before, I had been in New York. The glittering heart of Madison Avenue had revealed a similar world of wealth and status. There I had gone to the sumptuous shop of the Italian gun makers Beretta. They had chosen to have their flagship store in New York, not Milan, but stepping through its heavy entrance door, framed by a hand-cut Italian stone façade, you were immediately transported into a European world of precise luxury. The ground floor was devoted to thick shirts and jackets that were a frenzy of buckles and pockets. The top floor was filled with shotguns whose price tags you had to look at twice to make sure you had not misread them. But it was the middle floor, the walls filled with monochrome pictures of Africa and bookshelves heavy with coffee-table hunting books, that caught my attention. Because there stood a line of DVDs, and one leaped out. *Boddington on Cheetahs*, it read. But this was no David Attenborough–style film; rather it was highlights of the fastest animal on earth being taken down by a hunting rifle. Others stood beside it: *Boddington on Lions*, *Boddington on Leopards*.

What Boddington had done was strictly legal, but the images on the back cover felt like the sort of footage, as an investigative journalist, I would have wanted for a film about the ugly world of animal abuse. It seemed unnecessary and cruel. I am sure I'd be dismissed as a naive, city-dwelling liberal for this sentiment. Those who bought these DVDs would argue that there is no philosophical difference between shooting a boar or a cheetah; the latter was just nicer-looking.

But as I shook hands to say goodbye to Marc, I couldn't dispel the suspicion that perhaps he himself was not entirely convinced by the need to shoot lions and rhinos. I certainly wasn't. But, perhaps to challenge my own prejudices – if that was what they were – I had made a decision. After my visit to Cape Town, meeting medics and police squads, and before I flew to the United States, I would drive north, through South Africa's flower-lined Garden Route, to the remote frontier towns of the Eastern Cape.

The lodge lay about an hour from Cradock, a town of over 35,000 serving the farmers and traders of the districts that ran along the Great Fish River. Cradock had begun its life as a military outpost and was lined with neat, modest homes and wood-fronted shops selling the basics to live in this hard land. I had come here, to the western region of the Eastern Cape of South Africa, the poorest of all of South Africa's nine provinces, to go on a hunting safari.

Cradock was the last town before the Veld, empty but for the endless grass and, for me, animals to hunt. The town had the feeling of a place once filled with people who would have fought for something passionately. Today, though, it seemed those who lived here could not recall exactly what they would have fought for. It was a forgotten place. Perhaps a sense of self-belief died with the killings of four young South African activists here in 1985 – shot by white security police in the darkest days of apartheid. I had meant to see a memorial commemorating them, but the rain had come down hard from nowhere, submerging the view of the city. Instead, I edged through the crawling traffic until I found a sign that read Route 61 and pulled over to get my bearings.

The name places here spoke of a Boer past: Graaff-Reinet, Hofmeyr, Sterkstroom. This land, with its rolling, fertile expanse, the sky huge over it, was to the Afrikaner God's country. You could see why: here the winding light played across the plains, and the rain clouds could be seen for miles. I pulled out and drove on the empty road, a dark patch of sky approaching on the horizon.

The rain came harder this time, and then suddenly, through the water, a sign appeared: 'Fish River'. I took its lure. Ten kilometres and eight cattle-grids later I reached my destination.

It had taken me nearly three days of solid driving to get here from Cape Town. But distance is a feature of all life here. The hunting ground for Richard Holmes Safari stretched for some 200,000 acres on either side, bordering the edges of the Karoo and Eastern Grasslands. And after much negotiation about hunting permits and emails that bounced back and forth, I had agreed that I would come to these spreading valleys and gentle hills to hunt two springbok, antelope-gazelles, the national symbol of South Africa. This was no easy decision. In a very basic sense I was coming to hunt purely for this book. In order to understand the hunter's allure, I felt I had to do this. I had but one caveat – that whatever was hunted would grace the table later.

Getting out of the car, I looked around. I was in a shallow valley, and far away the tips of distant mountain ranges could be seen. Glossy starlings lined the road, high on branches that still dripped from the downpour, and the air smelled clean. And then, through the slanting rain, the owner of the Safari Lodge came out to greet me.

Richard Holmes was not who I had expected. Perhaps I thought he'd have a *Henderson the Rain King* spirit to him, a man with endless tales of derring-do. But instead he was considered and measured and had more the appearance of an accountant than a hunter. Of course, hunting was in his blood – he didn't have to dress for the part. He was seven years old when he shot his first springbok and had taken down birds three years before even that. He had run safaris now for over two decades.

In that way he was like many others throughout this country. At the turn of the twenty-first century there were almost no game farms in South Africa. Today there are over 12,000 of them, with 10,000 permitting hunting.[27] It's a big thing for these remote economies. The hunting industry generated 7.7 billion Rand – about $800 million – in 2011, with a third of that from the 15,000 trophy hunters who came here from overseas.[28] A lion hunt can cost up to $70,000, and a permit to hunt a black rhino recently raised $350,000 at auction.[29]

But, unlike some others, Richard only hunted free-range animals that he could eat; he would not shoot a lion or a cheetah for its pelt and had only ever hunted in South Africa. His wife, Marion, had the same approach. They also ran a conservation trust from their lodge for servals, caracals, African wild cats and black-footed cats.

The lodge was modest and neat. A section for butchering and refrigerating stood beside the Holmes's house, and beside that were a few thatched huts for the shooting guests. To one side was a flower-lined garden, filled with red and yellow blazes, the purple starlets of the African lily and the tight orange heads of gerbera daisies: an oasis of contained beauty in this wilderness. Further along was a dining room and kitchen. There was a filled fridge with vodkas and tonics, and outside was a fire-pit for a braai – a South African barbeque. I was told a dinner of game meat and vegetables would be served at dusk and was left to fall exhausted onto my bed.

That evening I went out to meet the other guests. There was one couple here: John and Doris White. John was a big and bluff American in his mid fifties, Doris fifteen years his junior. She was from Córdoba in Argentina, and he was from Minnesota – he had an identical twin who had also married an Argentinian. This he told me within a minute of meeting me, because like many Americans he was generous with his facts. I liked him. They had already paid for their hunt – four springboks, two blessboks and three ostriches – and both their faces were flushed with excitement at being here.

John had been raised with guns, like many American hunters I had met. He had been taught to shoot a .22 rifle by his grandfather, but years in the American Air Force flying Lockheed C-130s out of Panama, Spain and Greenland meant he had only recently returned to this passion. Now, with Doris, he was set upon taking down an ostrich – a hell of a bird to get up close to. The two of them, in matching camouflage T-shirts and trousers, were in hunting heaven.

The other guest that night was Jéane Grieve. He was a local taxi-dermist – in the trade for ten years after quitting his job as an aircraft engineer. American hunters like John made up almost all of Jéane's clients, and that explained why he was here – to drum up some more trade. Americans like to commemorate their hunts. Between 1999 and 2008, two-thirds of the 5,663 lions killed in Africa ended up being shipped out to the US.[30] And when it came to their trophies, the Americans wanted them big.

'They like full body mounts,' Jéane said. 'The rest of the world, except possibly the Australians, prefer bleached skulls.' I had never met a taxidermist before and knew nothing of his art. So I asked which was the hardest animal to work on. He was quick to answer.

'The porcupine. A full mount would be one of the hardest; the skin's paper-thin, particularly around its backside. It's a real pain.' Sometimes the issue is not the detail, but the time it takes to prepare an animal, he said. Elephants take two years to mount – his company has to outsource the task to others, they are so big. His own outfit has a warehouse with thousands of carcasses in it that are being treated. One client had 146 mounts ordered in one go. Many others want to have the big five mounted – the lion, elephant, Cape buffalo, leopard and rhinoceros.[31]

I looked a little shocked that anyone would want all five in their home, but he had a straightforward attitude to the animals that were hunted. 'I don't see a difference between hunting a kudu or a leopard, as long as it is properly managed.'

Hot plates of meat and potatoes were brought through, and we sat down at a long table to a dinner of hearty red Cape wine and freshly hunted springbok – imbued with a taste so far from farmed meat you cannot call them the same thing. The conversation turned, naturally, to hunting.

Each flowing idea was to some degree bold and, to an urban creature like me, previously unconsidered. Why not have a rhino farm where you could shave the horn annually, like shearing a sheep? Condemning the hunting of rare animals is myopic. In any pride there would be ageing females and males no longer fit for breeding, so why not license those for the hunt? 'They'll die anyway.'

'Human beings have caused the decline of predators,' said Richard. 'Now you have wild animals that need to be controlled. They need to be killed anyway. What is the difference between natural culling and a trophy hunter? You can't put a wild springbok in an abattoir anyway. By the time you did, if you could, you would stress them out so much that you'd have to tranquillise them.'

He said that their reserve helps protect habitats for wildlife, on land that would otherwise be turned over to agriculture. 'Sheep and cows create a farming monoculture of grazing that is deeply destructive.' Safaris have revitalised the land. In the 1950s, there were 500,000 game animals on South Africa's plains. Today there are 20 million, bred for hunting and conservation.[32] The gun did that, he said.

What about guns, I asked. Didn't all these guns increase the likelihood of people being shot?

'People kill people,' Richard replied. 'Guns don't kill people.' And, from his perspective I could see why he might believe this – farm murders are few and far between out here. But I had not come here to talk about gun murders – at least, not the killing of people. Besides, it was late, and I was due to get up before the dawn.

The plains were laid out below us, and above us thick clouds that never seemed to rain cast long shadows across the stretching emptiness. The wind was light, pushing the rough grass with jagged jerks. I clutched my rifle. Through my telescopic sight I picked out distant antelope in the quivering crosshairs, but they were too far away. We had been out for two hours now and nothing yet.

Then, John Sihelegu, my guide, touched my shoulder and pointed to the right. There, above a rocky gully, stood a springbok female. Perhaps seven years old, she had not seen us and was high up, framed against the cobalt blue. Only her top half was visible. Fine thatching grass covered her legs and belly.

I turned and raised my rifle. It was an awkward shot; the red rocks behind me dug into my kidneys. I felt the smooth wood of the Finnish rifle's stock on my face and closed my left eye. The sights lined up, and the crosshairs pushed down to the spot where I had been shown to aim, the best place to take her down: the heart, just beneath her shoulder.

My finger caressed the trigger, and I breathed in.

The shot rang out, and she fell. Then there was a flurry of noise as the staccato report carried across the plains and a push of pounding hooves as the herd rushed down the gully before us, leaping over the ochre rocks in their fear and confusion. Five, six, seven young springbok flew in front, muscles taut. They sped out down into the plain. I reloaded.

All had fled down, except the one I had shot. She lay above, unseen. Then I noticed that a small springbok, the last down to the gully's exit, had stopped, her tail quivering. She turned and looked for a second back up the ravine. She was turning for her mother and then she too was gone, bounding after the others. I had a terrible, lurching feeling in my stomach.

I stood up, gripped the rifle and, pushing away from the jutting stones, turned up the hill. The springbok lay there, twitching. She was not dead. I hurried to her and she panicked through the pain and the sweat. She tried to run but she could not; her shoulder was no longer there, and she could not stand.

The guide said, 'Take her. Quick. Put the sights to three and aim just there.'

I twisted the scope back from 6: 5 . . . 4 . . . 3. Then raised the rifle. I was five feet away and picked out the spot and again pulled the trigger. The springbok convulsed, and a small red dot appeared. Then she lay still, and that was it.

I felt nothing but sadness.

The guide told me to pick up the dead animal. 'Time for a photo,' he said. A large string of bloody snot was oozing from the beast's nose. Her muscles quivered, and her eyes quickly glazed. I picked up the warm body and pulled her onto a termite mound. And in the flipping, saw what the bullet had done. It had left a small entry point, but the back of her shoulder had been blown clean open. A hole, a deep cavity was there, just below her spine, and all the bones and muscles were exposed in an open, bloody mass.

I shifted the antelope's hind legs and placed it on the rust red earth. Then the guide told me to hold its neck. He wanted me to position it so that the best picture was possible.

'That's right. Just there. Take off your hat,' he said. The muscles in the animal's neck contracted. The picture was a good one. The scudding clouds behind me were full. The colours were vibrant, and the animal looked dignified. But my face in the photo was not one I had seen before: my eyes looked like the eyes of a killer, dark and full of bloodlust.

I stood up, and we got to work on the felled animal. A slickened knife opened up its gut, and its entrails spread in a slurry over the stony ground. Then it was lifted and hoisted down off the peak, its head lolling to the side, and with each step down I thought of the sharp crack of the rifle and the calf that looked back for its mother, and I wondered what this journey was doing to me.

11. THE SEX PISTOLS

Interview with 'God', a porn starlet called Stoya, in Las Vegas, USA – sex and guns – things seen in a different way – a Brazilian gangster, armed with his gun and his ego – men as victims not violators – the Pakistan conundrum – the persecuted journalist and the male logic of being armed and dangerous – talking self-defence with a mother of six in Washington, DC, USA – a lawyer's interpretation of statistics – asking why liberals don't talk much about guns in New York City

Perhaps it is not that surprising that at the same time the Shot Show – the largest gun show on earth – is happening in Las Vegas, so too is the AVN Adult Entertainment Expo, the self-proclaimed 'World's Largest Adult Trade Event'.

I had been to both. The two audiences were not dissimilar: single white guys. They shuffled, were modestly overweight, had unexplained stains on their checked shirts and sported beards. They were also very passionate about the subject matter. And 30,000 visitors to the latter of the two shows had been drawn towards this desert city in Nevada by the whiff of sex, making a lonely pilgrimage to gawk and take photographs.

There were other similarities. Like the Shot Show, the adult expo had the latest in accessories, each taking a basic concept to its logical consumer extreme. There was a dildo that you could strap onto your feet; vibrators that connected to your phone; and masturbating toys called Fleshlights made from moulds of porn starlets' genitals.

But it was not the allure of bullet vibrators that brought me to a sex convention. I was here for Stoya – a raven-haired adult entertainer and the queen of alternative porn. I wanted to speak to her because she had – for reasons unknown – appeared as a sniper in one of her latest films.

I had first heard of Stoya on an arts website. She had starred in a series of films called 'Hysterical Literature'. Its conceit was simple: a camera had filmed some of New York's most liberal women reading from works of fiction while, off screen, they were pleasured with a high-speed vibrator. The camera whirled as each woman was brought to a climax. Stoya had orgasmed while reading aloud about death, from *Necrophilia Variations* – a monograph on the erotic attraction to corpses. It was an intriguing choice of fiction, and so I looked up her name and found that Stoya, as well as being a muse on New York's art scene, also had sex on camera for a living.

She had joined the US porn scene at a unique moment. She was not a typical Barbie starlet: she wore Vivienne Westwood and had short hair. In her own words, she was gamine with small breasts. But clearly she had something about her, and awards for her sex scenes soon followed. In 2008 she won 'Best US Newcomer'; in 2009 she won a gong for the 'Best All-Girl Group Sex Scene'. But it was her role in *Code of Honor*, for which, in 2014, she won the 'Best Scene in a Feature Film', that intrigued me, because Stoya had been cast as a sniper, a character called 'God'.

So here I was, in Vegas, waiting to interview Stoya, because I wanted to know why playing a sniper was at all sexy. I had no idea. And in a sense I thought that 'God' could tell me.

She had asked to meet me outside the Expo next to a restaurant called the Pink Taco in the Hard Rock Casino. I arrived early, sat down in the soul-sucking light that permeates all of Vegas's casino floors and watched a series of enhanced blondes totter past on impossibly high heels.

Before our meeting I'd done some research and seen that Stoya had starred in another series called 'Stoya Does Everything'. On first reading the title, I had imagined illicit pleasures, but it turned out the films were for die-hard Stoya fans who wanted to see her doing stuff that did not actually involve her getting naked – things like pinball, ghost hunting, cosplay. In one she had visited a gun range.

The clip began with her line: 'Guns make me very nervous.' Then – a heavy-metal soundtrack kicking in – she was shown putting on a bulletproof vest and holding up a T-shirt that said 'Zombie Repellent'. She then headed down to the shooting range, and a man in military fatigues showed her a pistol. 'Big', she mouthed, her fingers outstretched.

She opted to shoot the smallest gun there – the Mosquito.[1] She seemed anxious; it was hard to tell if she was acting or not. But with her second shot, a projected shell casing spun out, knocked off the side of the gun range partition and hit her: 'Son of a bitch!' She left the range upset. 'There's a shell,' she said, pointing at her breasts, 'in my shirt.' Later, dragging on a cigarette, she turned to the camera: 'Now I am all freaked out and I don't want to touch guns again.'

Stoya arrived at the coffee shop in the flesh. She was louchely smoking Parliament cigarettes and wore a demure smile, a safety pin in her ear and a black corset on her slim frame. It was all very porn chic. But she spoke softly and with consideration, and it was clear she knew how to put someone at ease. She showed me her Vivienne Westwood get-up ('I live in New York and keep my shoes in LA'), talked a little bit about her cats and told me how she had a secret horror of Las Vegas ('five miles of death'). It felt entirely normal to be in a badly lit coffee shop interviewing her next to a porn convention.

I explained what I was doing. As she listened she took small sips from a coffee that had been served in a wine glass and considered her reply. 'One of the things about adult entertainment,' she said, 'is that it has to be inherently superficial. I mean, what do you think of when you have an orgasm?'

I spluttered. I'd never been asked that in an interview. She answered for me. 'Oblivion. Nothingness. Don't you? At that moment you are probably not contemplating anything. So in a film if you are trying to get across deep philosophical concepts you are pretty much shooting yourself in the foot.' She was telling me not to look too deeply into what role she had played; it was a porn film, stupid.

'When you do an adult feature you are making a film for about fifteen minutes in terms of the plot, and the rest is sex,' she explained. 'And so you have to work in archetypes in those short sections. So you have cheerleaders, nurses, moms baking apple pie. And one of those things that Robby D [her director] is good at is making chicks with guns look awesome. So playing a sniper had nothing to do with me. That was down to Robby D.'

There were certainly sexual archetypes that I had heard of before that focused on female snipers: black propaganda all. Russian forces in Chechnya in the 1980s spoke of blonde Amazonian biathletes fighting against them as sniper mercenaries; they called them 'white tights'.[2] The Russians sighted them again in South Ossetia and Georgia in 2008, and in Ukraine in 2014.[3] Then there was the female Viet Cong sniper known as Apache, who they said used to torture US Marines, letting them bleed to death.[4] One of her 'trademarks' was to castrate her captives. Later I learned that Dr Ruth, the famous American-Jewish psychosocial sex therapist, had also once trained as an Israeli sniper.[5]

This was the subverted image of the markswoman: sex, death and fear all wrapped up together. The erotic power of the unseen female killer – a shadow figure called God – bringing death and sex to men. It reminded me of Mexico's dark goddess of death that the gangs pray to – Santa Muerte – or the Israeli sniper I had met. Certainly I had been struck by how attractive the latter was, as if, somehow, her beauty, in an unseemly way, was incongruous with her ability to take life so easily. Perhaps this allure was what Robby D was seeking by casting Stoya in this role.

The reality, though, was quite different.

'I mean, I don't even like guns,' Stoya said, lighting up another cigarette. 'They frighten me.'

There's certainly a link between sex and guns.

On a superficial level you can point to things like videos of women in bikinis shooting rifles. This genre was most famously captured by Andy Sidari in his vacuous series 'Bullets, Bombs, and Babes' in the late 1980s. It featured a montage of unreconstructed sexual clichés – Playboy Playmates and Penthouse Pets wielding semi-automatics in G-strings. Men's violent fantasies made flesh.

More recently, as with Stoya, the gun has entered the peripheries of hard porn. The 'teenage starlet' Indigo Augustine seems to have built up a loyal following partly because her breasts are tattooed with inked images of pistols. And there are plenty of videos on porn sites with titles such as *Chick Has a Gun to Sell, I Got a Gun to Suck* or *French Girls Shoot Guns Then Have Orgy*. This fetish clearly says more about the masturbators than about the women involved. But seeing things through the prism of sex and gender helps in other ways. It makes it slightly easier to peer through the looking glass to see what guns really mean to men.

Of all the people I'd met on my travels so far, only a few had been women, and even fewer had been holding a gun in their hands. There was the doctor in South Africa, the funeral-home saleswoman in Honduras and the suicide-charity worker in Switzerland. But all of these women, in one way or another, were responding to the changes the gun had wrought; they were not the ones wielding them.

The ones holding the guns were almost always men. This did not surprise me; the odds of a man in the US owning a gun are at least three times higher than for a woman.[6] In Portugal it's even more extreme: 99 per cent of firearm licence applications are from men.[7]

Why is this? Historical influences and the male bias of many of the jobs that require guns offer some explanations. But it might also be a response to something else. After all, as female emancipation and sexual equality has edged its way forward, it seems that guns still remain as unreconstructed an instrument of male violence as they were 100 years ago.

This is possibly because in many countries modern market economics have caused the manly virtues of muscle and brawn to be challenged by the more 'feminine' charms of intellect, creativity and guile. Putting it simply, in developed knowledge economies, IT computer programmers and public-relations executives earn far more than construction workers and security guards.[8] This shift in desired skills has happened alongside a major manufacturing output decline, particularly in the US. There, nearly 30 per cent of all jobs in 1960 were blue-collar. Today it is about 10 per cent. Despite a near doubling of the American population, there are now about four million fewer manufacturing jobs there than fifty years ago.[9]

The outcome of this decline has been well documented: a pervasive insecurity about a man's role in the world (even if this insecurity is often based more on the perception of lost power than is actually the case – men still earn on average more than women).[10] And, for those men who feel their status has been eroded and confused by an ever-changing world, the gun is one thing they can cling on to. It redresses their sense of manliness and imparts the illusion of strength, status and power.

It's not me coming to this conclusion: the gun manufacturers themselves blatantly acknowledge this zeitgeist. Their adverts are full of vitalised masculinity. Glock says its guns give you the 'confidence to live your life'.[11] 'Balance the Power in your hands today!' is how the Tavor Semi-Automatic Rifle ad tells it.[12] The Walther PPX pistol is just 'Tough. Very Tough'.[13]

The manufacturers of the Bushmaster automatic rifle pushed all of this to its logical extreme. They ran a highly successful 'Man Card' campaign, where owning one 'confirms that you are a Man's Man, the last of a dying breed, with all the rights and privileges duly afforded'. Your friends can even revoke your man card if you are a 'crybaby' or 'coward'.[14]

The Small Arms Survey have even added their august views to this issue. They noted that men often turn to guns to assert their place in society: 'They are powerful tools with which young men can assert their masculinity, whether by acquiring the objects and status they are conditioned to seek, or by overturning the societies from which they are excluded.'[15]

All of this, to me, made sense, not least because I had witnessed the reality close up years before, a few days after seeing that murdered woman in the early hours of a Brazilian morning.

We had agreed to meet in one of the squat, square homes that fill the dead spaces and shantytowns of São Paolo's outer reaches. It was a district where the houses look like children's drawings – places thrown together and hemmed in with corrugated iron and breeze-block. He had told us not to be late, so we waited for him inside, behind a high metal door that felt like a prison's gate. When he walked in, a thin and tight-wired man of nerves and reptilian movements, he greeted us wordlessly with an elaborate handshake and then looked around the place with a suspicious eye. He had a beany hat pulled low, despite the heat.

Pulling up a plastic chair, he motioned for us to sit at an unstable table. Perhaps this piece of furniture served as an object of respectability. It made this seem more like an interview than a snatched conversation that broke the law. He was nervous, and tension crackled in the air. Then he reached under his shirt and pulled out the thing we had come to see – his pistol. He showed it to us slyly at first, and we leaned in to get a better look. It was a US-made gun – a .38 Harrington & Richardson, made by an American company that now operated under the Freedom Group umbrella of gun companies. It must have been over thirty years old.

I had wanted this for the film I was making: to meet a gang member with an illegal gun. Oddly, in El Salvador and Honduras gangland guns had been kept hidden from me, but I had got what I sought here.

'I carry this,' he said, his arms strong and sinewy and marked with a history of violence, 'because of the power it gives me. It makes me feel like I am taller. Like I am bigger.'

The reporter – Ramita Navai – asked if he had killed anyone with it, and the translator said it was better if we did not ask that question. Instead he carried on talking: 'Here – this can make you someone. The police, other gangs, other men – they listen to you. This one talks.' And he held it out again – a heavy five-shooter with a worn edge and a dull patina. He spoke in that slow singsong way. With all its cadences and rhythm, Brazilian Portuguese can sound like a balmy evening on a beach with a cold beer. But right now it sounded like a death threat, like a serpent's hiss.

The way the gun had the capability to make this slight, rat-faced gang member feel like a tough guy is something that exists across time and culture. It has even led Todd Hartley, a columnist in Colorado, to suggest that the best form of gun control would be a campaign to challenge masculinity at its roots, to insinuate that men who bought semi-automatic rifles had small penises.[16] This might sound ludicrous, but here in Brazil a peace charity, Viva Rio, did pretty much that. As a lobbyist for them, Professor Antonio Bandeira, said in an interview: 'One of our most successful media campaigns ironically associated sexual insecurity with the glorification of guns. Pretty and popular actresses said, "Good lovers don't need a gun."' The campaign helped the charity lobby for tighter Brazilian gun control, and, as the professor said: 'Research showed gun homicides declined 8 per cent in the five years after the law was implemented in 2004, saving 5,000 lives.'[17]

I think Viva Rio understood something that many gun control lobbyists elsewhere fail to. That simply imposing gun laws without a cultural shift in attitudes to guns is not going to work; that you have to look at hard psychological truths. If you view attitudes to guns through the lens of sex and desire you might see, to take inspiration from the French psychoanalyst Jacques Lacan, that man's desire to own a phallic symbol like the gun is a response to something deeper. A shrink might argue that gun ownership is a response to the fear of emasculation – a reaction to metrosexual modernism where men face losing their jobs, their status and, ultimately, their meaning.[18]

Any threat – such as the possibility of men losing their gun – causes, in turn, a deeper fear. This might explain why even the most basic gun legislation is met with fierce opposition in the US.

You see it in the details. Like when a Florida gun instructor put a video on the Facebook page of the pro-gun control group Moms Demand Action. In it, he shot a target bearing the group's logo. 'Happy Mother's Day,' he said, showing the bullet holes to the camera.

Or when forty members of the group Open Carry Texas – who call Moms Demand Action 'thugs with jugs' – showed up outside a restaurant in Arlington, Texas, armed with semi-automatic rifles. Four women from Moms Demand Action were inside eating, and the men stood outside posturing in a bewildering way. Members from that same armed group were later to hold a 'mad minute' at a firing range, pulverising a female mannequin with a hail of bullets.

These small events are so ugly as to make you scratch your head. But psychology perhaps helps us understand why such intense, disturbing emotions might come about. It shows how the debate is not just about guns but about other, deeper fears.

Looking at gun ownership in psychological terms also did something else for me. It helped me challenge the notion that 'guns don't kill people, people kill people', as Richard had claimed back in that hunting lodge in South Africa.

In 2006, psychologists at Knox College in Galesburg carried out an experiment on thirty male college students. They each provided a saliva sample and then were given a gun or a child's toy to play with for fifteen minutes. Afterwards another sample of their saliva was taken. The men were then asked to add as much hot sauce as they wanted to a cup of water they thought another subject was going to drink. Those who had been given the gun showed significantly greater increases in testosterone and added much more hot sauce to the water than those who had played with the toy.[19] In this experiment the gun seemed transformative. Other studies have shown similar results. Carrying a concealed gun has even been said to change the way people are seen to walk.[20]

Of course, such experiments need to be replicated a dozen times to prove anything. But I can say this: I have seen policemen and soldiers, self-defence trainers and criminal gangbangers visibly change when they picked up a gun. I have watched the way their guns became part of them, the barrel an extension of their bodies and wills, emboldening them and diminishing the rest of us who did not have guns. The gun transformed these men and the entire situation with it, just as it changed the eyes of the thief who sat opposite me in the heat of that São Paolo summer.

Afterwards, as we sat in the crew car and drove away from the dusty favela, our translator told us she had been terrified. The gang member had spoken words that oscillated between paranoia and anger, and she had struggled to comprehend some of the things he was saying. And she had feared the small pistol in the hands of this angry man.

But this is what happens. A gun gives that ultimate edge of authority to someone who lacks it through intelligence alone. On its own the gun wins any argument – it elevates 'A Nobody' to 'The Man'. Small wonder so many men love them.

As with so many love affairs, things quickly go sour. So it is with man's love for the gun – a love affair infused with death.

With the exception of New Zealand, Hong Kong, Japan, Korea, Tonga and Latvia, more men are killed as a result of armed violence than women. Globally, the male homicide rate is almost four times that of females.[21] And in places such as Brazil the vast majority of homicide victims, over 90 per cent, are men – usually poor and young ones.[22] For every woman killed by armed violence, the World Health Organization has reported that thirteen men are killed in Colombia, fifteen in El Salvador, sixteen in the Philippines and almost seventeen in Venezuela.[23]

In war, men fare similarly badly. A major analysis of global mortality data found that the male deaths in war always outstrip female deaths. Also, the females with the highest death rates are baby girls; in males it's those between the ages of fifteen and twenty-nine who are hardest hit. This is because women are often impacted by the indirect consequences of war – disease and malnutrition. Men are impacted directly by gunfire – herds of young men going to their death.[24]

As I had found out from the World Health Organization, men also commit suicide more often than women. The extra-governmental agency looked into gun suicides and found that in every country they reviewed men had killed themselves with greater frequency than women.[25]

Yet these terrible and sad figures about the impact of guns on men are often not debated with any conviction. It's almost as if we feel this is all inevitable, as if men somehow deserve their painful and lonely deaths at the hand of a gun.

Google the exact phrase 'Armed violence against women' and you get 26,900 responses. Google 'Armed violence against men' and you'll get two hits.[26] Of course, men are overwhelmingly the perpetrators of gun violence, but this turning away from the true impact of gun violence on men has consequences that are deep and troubling.

For years, a spectre hung over Ciudad Juárez. In the 1990s, this northern Mexican border town became known for its gruesome femicides – the deaths of hundreds of women. These murders, often sexual in nature, grew in the public's imagination – referenced in Tori Amos's song 'Juárez', Roberto Bolaño's novel *2666* and FX's drama *The Bridge*. The city became synonymous with the rape and murder of young girls; pink crosses sprouting around its barren edges, each commemorating the bodies of women found in that grey Mexican dirt.

But as these female tragedies were unfolding in Juárez, far greater numbers of men were also being killed and mutilated. Between 2007 and 2012, over 11,400 people were murdered here. Despite having only 1 per cent of the Mexican population, this border town had about 9 per cent of Mexico's homicides. By 2010, ten people a day were being gunned down. That year over 3,500 people died, and Juárez earned the title of the most violent city in the world outside a war zone.[27]

And the vast majority of these dead were men. But few crosses were raised for them.[28] In fact, when you really look at the figures you'll be surprised. The *proportion* of women as victims of all homicides in Juárez was less than it is in many US cities. Female murder victims in Juárez between 1987 and 2007 worked out at less than 10 per cent of all killings there. In US cities such as Houston up to 20 per cent of those murdered in a given year are women.[29] And where women were murdered in Juárez, about three-quarters were the victims of terrible domestic violence – so the cases were essentially solved. Indeed, in those years, only about 100 of the murders of women there were of the type immortalised in film and print.[30]

Molly Molloy, a researcher into the drug violence that has plagued Mexico, pointed out in an interview with the *Texas Observer*: 'I've read things by some feminist scholars talking about the "harvest" of young, nubile women. I mean, the terminology becomes kind of sensual, or sexual. Some of the writing about these cases I find to be pushing over into the extreme and eroticizing the victims in a way that makes them appear a lot more helpless and powerless than women in Juárez are.'[31]

Her concern was that the murdered females of Juárez, in a place where about ten times as many men are killed as women, have somehow been overly focused on – to the point of being fetishised. Saying these hard truths does not diminish the fact that each and every death suffered by women there was terrible. But these truths should be said, because every single victim of gun violence matters, regardless of gender.

So why does it seem somehow, then, that men's deaths in Juárez are given less importance, less media focus, even less sympathy? It's as if we assume any man killed by a gun in Mexico somehow deserved it more than a woman. But assuming all men murdered there were narcos or gang members is ludicrous. There are the countless young Mexican men pulled off street corners to work as *halcones* (lookouts), their lives snuffed out just because they were in the wrong place at the wrong time and saw something they shouldn't have. Certainly as many of these 'innocent' men were killed in Juárez as women were. Focusing on the bodies of women as the victims in the bitter landscape of Juárez's violence serves only to address a small part of the overriding problem. As Molly said: 'If you look at the problem of violence in Juárez as essentially being a problem of young women being murdered, and that if you can solve those murders, then everything will be OK, it feels safer. It feels like you can accomplish something, because then you don't actually have to look at the real problems of the city.'

Ignore the killings of men, though, and you risk not finding any real way to address the violence committed by them and against them. You risk the malady of eternal violent repetition. And not just in Juárez – it is all over.

And yet . . . and yet. The answer to reducing gun harm is not as simple as just denying men access to them. After all, guns can be used as much in the preservation of life as in the taking of them. And the male urge to protect is, as I had seen on a trip to Asia, a very primal one.

I was to meet someone recommended to me. A true Pakistani gun enthusiast, I had been told. A man you have to meet – someone who has so, so many guns to protect himself and his family. Why, I had asked. Because the Pakistani state can't protect them. So I called, and he said, 'Yes. Come over.'

This was off the cuff. I was in Lahore mainly to talk about the levels of terrorist violence that had impacted this country of 180 million. There were rounds of press interviews and launches to attend on a campaign that my charity was running, seeking to high-light the plight of victims from the rising tide of suicide bombings. But being here, in a country so beset with murder and death, and one infused with machismo and paternalism, I felt compelled to speak to a man who owned guns, because the idea of armed self-defence in a nation where the need for it was so stark intrigued me.

Modern assault rifles and handguns had long ago come into vogue among middle-class male Pakistanis. They bought them because they had lost faith in the ability of the country's civilian government to protect them. They had seen the assassinations of its popular polit-ical leaders, a spreading Islamist insurgency and daily terrorist explo-sions, and so they prepared themselves for the worst. But more than this, Pakistanis feared the hidden threat of violent kidnapping, extortion and robbery. The murder rate here was 53 per cent higher than in the US.[32] And the man I had agreed to meet was more exposed than most: he was a Pakistani Christian, a father and a journalist, all of which brought with them their own challenges.

So, long after dusk had fallen and with the headlights of my car piercing the corners of a crowded suburb of Lahore, I travelled with a local guide to meet Mr Asher John, the chief news editor of *Pakistan Today*. The mad rush of Lahore's busier streets, with its ramshackle shops and pulsing electric lights, gradually eased off, and we entered a quieter section of shaded corners and suburban calm. Security guards, wrapped in thick shawls to keep out the night chill, stared from the gloom. Then the car beams picked out the nameplate on his home: John's Lodge. After a beeped horn, he came outside.

In his mid thirties, Asher was trim and balding and wore a neat moustache. He looked like a typical father, with his jeans, checked shirt and sandals, and he ushered me into his modest home with the mild insistence that comes with Pakistani hospitality. There, underneath a mounted display of a deer's head, he offered me a Coke and sat and waited for me to ask him a question.

I began by wondering why he had developed such a reputation as a man who knew his guns.

'If someone has a gun and is trying to kill you, it would be reasonable to shoot back with your own gun,' he answered, in the peripheral way that many Pakistanis do. The low light caused by weak electrical power, one that marks the nights of this region, gave the conversation a conspiratorial edge.

'Gun violence in Pakistan is pretty common, you can get shot in the rural areas, you can get shot on the roads. In Pakistan, we're used to seeing people getting shot, and we've ceased really caring about it. Beheadings are something not common: that makes news. People getting shot is everyday.'

He had more to fear than most, though. At thirty-six, he had been a journalist for thirteen years, and his beat was a dangerous one. He mainly reported on blasphemy cases, especially where guns were used in attacks on minorities, including Christians like him.

'There was one time, five months ago, I was being followed from my work,' he said. 'We were trying to investigate a Muslim cleric who was raping young boys at a seminary, and these thugs came after us. They said we made the story all up. It was easier for them to say we are trying to malign the cleric, to call us blasphemers, even though we had eight or nine children on record. I had to fire two shots to scare those people away.'

The lights from outside flared over the high brick walls, and the grilled windows cast prison-bar shadows on the ceiling. I tried to imagine – from a man's perspective – what it must be like to live with your family in this way, as if there was nothing solid underneath to guarantee their protection. I guess, for him, guns were the base that he stood on, the thing that made him the protector that he wanted to be.

He brought down a .44 rifle that had been converted into a fully automatic assault rifle here in Pakistan, and handed it over. 'I don't know how to make this safe. How do you make this safe?' I said.

His phone rang, and he said casually, 'It's safe.'

But it wasn't. My right hand cocked it to see if there were any bullets in the chamber, but in this dim light I could not see clearly inside, and as I let the cocking lever go the noise of a round being chambered caught his attention.

'Oh. This one's loaded,' he said suddenly. Another second and I could have fired a full magazine into his living-room ceiling. I guessed, too, that if you needed a gun for self-protection then there was no point in not having it loaded. He laughed and carried on as if nothing had happened.

'It's not just a hobby. It's more of a necessity now. I will train my daughter how to use a rifle, given that we have weapons and we are the only Christians who live in this area. It's important for each of us to know how to fire a weapon.' He talked about the violence that marked his village, and how he had lost cousins to a thirty-year-old blood feud over property rights that still haunted his family out in the provinces. It was a disagreement that had always been framed by the presence of guns.

He had fired his first gun, his grandfather's shotgun, when he was just ten years old and had bought his first weapon when he turned eighteen: a .30 calibre pistol for protection. Since then his gun rack has felt the weight of new purchases. Today, he has about twenty guns – the number alone a display of strength.

'A local-made AK47 costs around $600 American,' he said. 'A

Russian one would cost around $1,500 and a Chinese one $1,250. There are foreign [allied] weapons, too. These are usually guns that have been stolen from the troops in Afghanistan. There are a lot of weapons coming in from there: M16s, NATO-manufactured AK47s, AK56s. The most commonly used guns we have are local-made ones, though: .34 calibre pistols. You can get anything in Pakistan.'

Craft production – a small-scale, hand-lathed industry – was a major source of guns in the northern tribal regions of Pakistan; 20,000 weapons were produced every year.[33] It was a double-edged sword – these arms both offered protection and posed a permanent threat.

This man's dilemma, then, was the eternal joker in the pack of gun control. In a country where extra-judicial killings are well reported and where there are high levels of gun violence, there is a philosophical justification for arming yourself in order to protect your family. Who could deny this articulate, humane man his right to defend his home? When police forces are corrupt and religious bigotry means that masked men follow you home, the gun becomes more than just a hobby. It becomes a means of survival.

The trouble, of course, is when this argument for self-protection is used in societies that are fully functioning – in places where there is a trustworthy and well-equipped police force. Then the right to defend yourself arguably needs to be weighed up against the logic that you should undergo rigorous vetting before you can own a gun. But for this man – an infidel in a land of devout Muslims, a journalist in a country where telling the truth can get you killed, a father in a culture where honour killings and femicide are rife[34] – having a gun seemed the only choice left. At least to do what a man often feels he has to do – to defend his family. It was either that or leave the country – something he wasn't prepared to do.

This said, though, what happens if you live in a seemingly orderly society and yet still doubt your safety – like those Americans who think their pistols and revolvers are the only thing that can guarantee their security?

To answer this, I felt that I had to look at the issue of women and self-defence. And so, a few weeks after returning from Pakistan,

A midnight death in San Pedro Sula, Honduras

Dark sacks in San Pedro's morgue (*above*)

Death graffiti, San Salvador (*right*)

Children searched for drugs and guns in South Africa

Spot the Israeli enemy

Clay shooting in Iceland, a country with one of the lowest homicide rates

Asher John's home security in Pakistan

Handmade luxury at the IWA show in Nuremberg, Germany

Meeting Stoya in Las Vegas

Sex sells at a gun show

Bullets and burgers on Vegas' strip

Gambling for guns at the Shot Show

Satanic imagery at the world's largest gun show

Poor attempts at gun humour

Kalashnikov and his legacy at the Las Vegas
Shot Show

Glock's gun stand

The enemy is everywhere – the Eurosatory
military show in Paris

Underwater guns from the Russians

Being shown around the Contraband Museum in Odessa

No More Weapons, Ciudad Juárez

Pigeonholes filled with new pistols in Samsun, Turkey

Zombie menace at the Orange County Fairgrounds
– the perfect marketing tool

I found myself travelling back to the United States. This time I was to head to the heart of American politics – Washington, DC – there to meet a pro-gun advocate who believed that gun control itself was sexist.

I had arrived at the café early in west Washington, DC and was nursing a hot drink and the beginnings of a cold. Outside, Georgetown's neat streets pushed out and beyond. This was M Street, as if there were not enough politicians, orators and celebrities to name the streets after them ten times over. In this city, though, where everything is political, perhaps choosing names like this would be a conceit too far.

I was here to see Gayle Trotter. She was late, and I could not recall what she looked like from her online photos, so I approached a woman in dark glasses and asked if she was Gayle, and she reacted with wide, fearful eyes. I backed off and sat and listened to the conversation of the bald, bearded man five tables away.

Gayle was, to me, an important person to speak to when looking at guns through the sexual looking-glass because this mother of six had given testimony to the Senate about why the right to bear arms was a female right to self-defence. Hers was an interesting argument. When one senator questioned her, she shot back: 'You are a large man . . . a tall man. You are not a woman stuck in her house, not able to defend her children, not able to leave her child, not able to go seek safety.'

Then she argued that the most important thing about assault weapons for a woman was the way they look. That a big gun gives off the perception that the female holding it has the capacity to kill, rather than forcing her into killing.

'An assault weapon in the hands of a young woman defending her babies in her home becomes a defense weapon . . . knowing she has a scary-looking gun gives her more courage.'

It was, to some degree, the most convincing pro-gun argument I had heard yet. Given that women are so infrequently the perpetrators of gun violence, the idea of them protecting themselves from harm in this way seemed logical. It was even tempting to conceive of a form of gun control that only let women carry arms and not men – though clearly that would never fly.

Gayle arrived, apologetic. She was a trim and square-jawed woman in her early forties, born and bred in Washington and a lawyer now for eighteen years. This profession had been a calling, of sorts. When she was just eight, her lawyer father, a dominant man and a member of the National Rifle Association, had given her a copy of *Black's Law Dictionary*. In it he'd written: 'In the hope that you will see fit to pursue the law.' Clearly he was a man who got his way in life. He was now her practice partner.

Words spilled from Gayle in a torrent, so much so that it was difficult to decipher if it was nerves or just an incredible intellect at work. She made it clear, too clear even, that she was not a violent person and then, and I was not sure why, she told me twice she was not a vegetarian.

She was intrigued, too – why would I want to go hunting just for this book? 'But couldn't that be like the first time you have sex? It's not going to compare to someone who hunts their whole life.'

She had a lawyer's tone to her, always seeking a weakness in an argument, so I kept our conversation tight and asked her about why she had told the Senate that gun control, if it took guns from women, was a violation of women's rights.

'For me, guns assure that women can be free and equal,' she said. 'Without that you are relying on the government, or your husband, or whatever stronger man is around you, to provide that kind of protection. And, you know, women of an earlier generation were saying, "We're the same as men, and we need to take our rightful place in society," but sometimes those are the women pushing the strongest for gun control, and then they're ceding their own personal protection to the government or to stronger people who are in their area.'

She had a point about the need to consider self-defence in all of this. Even though gun violence against women is often disproportionately covered in the media, American women are still eleven times more likely to be murdered with a gun than women in any other developed country.[35] And little seems to be done to stop it – only nine states in the US prevent people jailed for stalking from, once released, purchasing a gun.[36] In 2014, it was estimated that there were almost 12,000 convicted stalkers in the US who were still permitted to carry guns under federal law.[37]

'Women should have the choice of what makes them feel comfortable to defend themselves, and there's no reason why you shouldn't have the ability to have a shotgun or an AR-15 or whatever variety it is,' she said. She had an issue, too, with background checks. 'If I were being threatened by someone, I wouldn't want to have to wait a nanosecond to be able to exercise my right to choose to defend myself.'

Her comments to the Senate had unleashed a storm of sorts. 'I got threats on Twitter, on Facebook . . . My personal favourite was: "I want to beat you to death with an X-Box."'

She had support, though. Her view was shared by other leading voices in the pro-gun movement. Wayne LaPierre, the executive vice president of the NRA and a man who shared her panel at the Capitol hearings, has said: 'The one thing a violent rapist deserves is a good woman with a gun.'[38]

There was a flaw, however, in Gayle's argument. Many studies have concluded that having a gun in the marital home puts women more at risk than men.[39] A national study on the murder of women in Brazil found 40 per cent of women in 2010 were killed at home, compared with just 15 per cent of male victims. It also found that 54 per cent of those women were killed with a firearm.[40] And in Canada women are about four times more likely to be killed by their spouses than are men.[41]

The situation in the US for women is as dire, if not worse. A global analysis of twenty-five high-income countries with populations over 2 million found that the US 'had the highest level of household firearm ownership and the highest female homicide rate'. It was described as an 'outlier'.[42] One study there found that a woman's chances of being killed by an abusive live-in partner was five times greater if that person had access to a gun.[43] As an American review of the scientific literature concluded: 'There is compelling evidence that a gun in the home is a risk factor for intimidation and for killing women in their homes.'[44]

It seems guns might well be the problem, not the solution. When it comes to 'stranger danger' an American woman is more likely to be raped in a gun-friendly state than in one with a higher level of gun control.[45] We know that, in the US, having a gun in the home increases the overall risk of someone in the home being murdered by 41 per cent. But what is shocking is the fact that, for women, the risk of death is actually tripled.[46]

Gayle's armed solution to protecting hearth and kin was one that made matters worse, not better. And yet, despite the statistical realities, the self-defence mantra dominates. Gun manufacturers sell women the idea that equipping themselves with a gun will make them a lot safer. When Smith & Wesson introduced the LadySmith gun they launched a feminist advertising campaign with it. 'Refuse to Be a Victim' the ads said – the appropriate language of empowerment through a handgun.[47]

You can now buy semi-automatic rifles in lipstick shades of pink; purchase concealed gun-carry 'designer' handbags and holsters in leopard-print and snakeskin; get women's lingerie that holds a pistol; and order novelty T-shirts (one reads 'Free Men Do Not Ask Permission to Bear Arms') from websites like www.gungoddess.com.

It is a sense of armed empowerment that is visible in the rise of all-female shooting clubs and courses, ones with names like Babes with Bullets and The Well Armed Woman; and a rise, too, of the proportion of all women owning a gun in the US: from 32 per cent in 2005 to 43 per cent in 2011.[48]

These increases are intriguing when you learn that, in 2010, the number of violent crimes in the US apparently dropped to its lowest rate in decades. 'The odds of being murdered or robbed are now less than half of what they were in the early 1990s,' the *New York Times* reported. 'Small towns, especially, are seeing far fewer murders. In cities with populations under 10,000, the number plunged by more than 25 per cent last year.'[49]

This, of course, can be read one of two ways. The first is that the arming of America's women has reduced the threat of violence to all. The second is that women are arming themselves against a threat that is itself declining. And in the space between these two conclusions lies the eternal debate that exists over the right to bear arms, regardless of what sex you are – a debate that is as much down to culture as it is to statistics.

New York's Museum of Modern Art was as good place as any to start. If there was going to be an edgy and artistic interpretation of the way that guns impact modern culture, a culture that reflects the use of guns, then surely MOMA was going to have it. At $25 a ticket, they'd better have, anyway.

That entry price certainly doesn't buy you much advice. Walking through the minimalist glass and steel interior, I made my way over to the information desks. They were manned by black-polo-necked volunteers and one, a camp man who told me he was from the Upper East Side, on being asked if there were any paintings in the museum with guns in them, suggested I should, perhaps, look it up on the internet.

But I didn't have any internet connection and so instead I decided to walk the galleries. I'd come here, on my way back from Washington, assuming they'd have the Lichtenstein *Takka Takka* image of a machine-gun blazing away, or Warhol's silkscreen Elvis Presley in a gunslinger pose.[50] But they had neither, and even their show 'American Modern', despite its endless Midwestern vistas and shady representations of street-corner cafés and secret losses, had no rifles or pistols.

So I paced the gallery floors, past students copying masterworks into sketchpads and young Japanese lovers in front of giant Rothko screens of green and blue, looking for images of firearms. Eventually I came across a sparse and threatening image called *Gun with Hand #1* by Vija Celmins.[51] And that, pretty much, was it. The title of Celmin's painting accurately described what it represented and – as with so much of the world of guns that I had seen – it allowed you to project your own prejudices and assumptions onto its starkness. But it was just one image in the centre of America's beating cultural heart.

Then it struck me – that the issue of the gun here in America, even though it claims over 30,000 lives a year – is largely ignored by the liberal coastal elites. You see it in America's mass media aplenty, but when it comes to opera, galleries and design museums, the gun is notable for its absence.[52]

It's not just MOMA. The Boston Museum of Art refuses to display firearms because they don't think guns are high art.[53] The Tate Modern in the UK has only three artworks that show a gun in its whole collection. London's Design Museum has only one – an AK47. And of the 200 or so firearms in New York's Metropolitan Museum's gun collection, none were made after 1900.[54] Those on display show great craftsmanship and decorative lustre, but they are not on show for their functionality. It's as if – in the curator's ideological view of the world – the modern gun is too ugly, too functional, too damn 'street'.

Of course there are reasons why this is so. There is the legacy of museums evolving out of philanthropic largesse – where collections were established on the basis of their beauty not their function. The safety implications for housing a gun in a collection might also be of relevance. As is the fact that many curators might think guns are things best found in armoury museums, not design museums.

Lurking within all of this, though, seems to be the concern that if you put a gun in a design or culture museum, as opposed to a military one, you somehow legitimise it. As Paola Antonelli, a senior curator at MOMA, said in an interview: 'Showing a firearm meant endorsing the firearm, endorsing its lethal power, endorsing its violent potential. It's about endorsing evil, in a way.'[55]

Yet, despite this curatorial control on elite tastes, the gun still has a huge influence on culture that cannot be denied. Gun metaphors long ago crept into our everyday chatter. Trends are called a 'flash in the pan' (what happens when gunpowder flares but the bullet doesn't fire); things 'fizzle' out (a muzzle-loading rifle 'fizzles' when it misfires); you 'bite the bullet' (wounded troops were once given bullets to put between their teeth as they underwent surgery). And others. You 'keep your powder dry', 'take potshots', use 'bullet' points.[56]

In terms of popular culture, the gun seems as ubiquitous. Of the top fifty films listed on the International Movie Database (IMDB) there are few that don't involve the use of guns in their plot. Levels of gun violence in mainstream American films have more than doubled since 1950, and the levels of gun violence in PG-13 rated films now outpace those of R-rated films.[57]

These films often tell a singular story, one suffused with an American view of popular culture that says: 'The only way to stop a bad guy with a gun is with a good guy with a gun.'[58] From Bruce Willis's John McClane to Clint Eastwood's Man With No Name, the American story is of the lone man forced to take matters into his hands, often with the aid of a gun. In the story, violence is usually regenerative: someone has to be shot so society can regain its peace and identity.

The trouble is that these ploys and plots have consequences.

Hollywood's influence on gun culture cannot be dismissed. When American celluloid gangsters used sawn-off shotguns, so did real-life gangsters; one study in four US states found that 51 per cent of incarcerated US juveniles once owned one.[59] When semi-automatic, assault-style rifles were used in *Rambo* and *Miami Vice* in the 1980s, sales took off in America. And I've seen gang members posing, holding their pistols sideways – a terrible way to aim a gun – and was not surprised to learn this was first done in the movies so that both the gun and the actor's face could be seen on camera simultaneously.

The dirty secret, though, is that the gun companies know all too well this Hollywood influence on sales and seek to build close relationships with prop houses.[60] Smith & Wesson, for instance, once reportedly hired a firm called International Promotions, a company that specialised in product placement, to push its firearms on the silver screen.[61] In 2010, Brandchannel, a website that spots product placement, gave Glock a lifetime achievement award after their guns appeared in over 15 per cent of that year's box-office-topping films.[62]

The gun's cultivated imposition on American mass media has far-reaching consequences. Researchers have repeatedly looked at what happens when children are exposed to violent media and concluded that it increases aggression and is a 'significant risk to health', something endorsed by six leading US health pediatric organisations.[63] The saturation of guns in films normalises them, reinforcing to many the belief that you need one to survive. And that, as another study found, 'the presence of weapons in films might amplify the effects of violent films on aggression'.[64]

It's not just films, though, that hold influence. When it was reported that Adam Lanza, the Sandy Hook mass shooter, had obsessively played the computer game *Call of Duty* in his basement, the media latched on to the influence of computer-game-inspired violence.[65] Certainly the popularity of 'first-person shooter games' is huge and, with their very realistic portrayal of mass gun death, can make gaming companies very tidy profits. *Call of Duty* brought in over half a billion dollars in just one day in 2013.

Other games have even seen brand partnerships that take entertainment into advertising; highlighted by videos such as those showing *Medal of Honor's* executive producer and a representative from the gun company Magpul looking at weapon accessories together.[66]

The link between violence and the influence of computer games, though, seems more nuanced than with film. The American Psychological Association concluded in one review that computer games have many positive effects on children; in some cases even violent games can have some beneficial impact. There are data to suggest games such as *Grand Theft Auto* or *Call of Duty* may have helped bring about a drop in crime levels.[67] Perhaps these games allow the taming of violent urges through play. Perhaps it's because games take up a huge amount of time, keeping young men from trouble. What we do know is that there is no hard proof that video games cause violence.[68]

What is also clear is that it's impossible to look properly at gun owners without seeing them through the prism of gender and culture, because sex and environment form the very ideologies that make people feel they need guns in the first place.

But where there is sex, there is also money. Profit follows pleasure, as night follows day. So I shifted my gaze once more and began to do what investigative journalists always do: follow the money.

V. Profit

12. THE TRADERS

Lords of War and global markets – the largest gun show on earth, Las Vegas, Nevada – selling patriotism and fear – conversations with AK47 dealers and arms traders – how guns end up in the hands of human rights abusers – France: Europe's largest arms fair in Paris – stonewalled by the Chinese, bullied by the Russians – Ukraine – shady arms ports in Oktyabrsk and intrigue in Odessa

The legal international trade in guns and ammunition is thought to be worth about $8.5 billion a year.[1] I was surprised when I read this. I had expected higher. But there it was: the sale of guns was less than 10 per cent of the conventional global arms trade, even though firearms can account for up to 90 per cent of casualties in conflicts.[2]

This figure also revealed another misconception on my part – possibly one formed by films such as *Lord of War*, where Nicholas Cage plays a debauched and corrupt US-Ukrainian arms dealer supplying Africa's wars with endless weapons. I had assumed many guns were smuggled illegally around the world. Not so. The vast majority of the trade – as much as nine-tenths – begins in legal transfers.[3]

Almost all guns start their lives in legally run factories, and such shipments of new guns have usually been stamped and approved by some ink-stained bureaucrat along the way. It's only later that they end up in a dirty supply chain run by smugglers and traffickers and pushed into ugly corners of violence and despair.

When compared to the global trade in drugs ($321 billion)[4] or people trafficking ($32 billion),[5] the legal trade in guns seems regulated and modest. At least until you see, as I had done, how devastating a gun in the wrong hands can be.

It's a market, though, that is growing. The global pistol and revolver trade more than tripled in value between 2003 and 2013.[6] As UN data shows, by 2013 almost 31 million firearms and parts were being traded by ninety-four countries worldwide.[7] In one year alone the US imported about $0.8 billion of guns from around the world and exported $0.4 billion.[8] This is just exports. Domestic sales add more guns, more millions of dollars. Over 98 million firearms were sold by Americans to Americans between 1986 and 2010.[9] Adding international and domestic markets together, it is clear that the US's influence on international trade in guns is not only dominant but a game changer. Without them, the world's relationship with the gun – and the numbers of guns – would be markedly different.

The simple fact is this: all the major gun companies rely on the Second Amendment to maintain profits. The right to bear arms is more than a matter of principle – it's serious business.

There are over 5,000 gun shows in America each year. The biggest by far, not only in the US but in the whole world, was Las Vegas's Shot Show.[10] Based at the Sands Expo Center, it was a behemoth of an event. It had grown and grown over the years until, bloated, it now spread over 635,000 square feet, with 1,600 exhibitors. It was a floor space that was the equivalent of walking into Terminal 5 at New York's Kennedy airport and finding it filled with guns. If there was one place under the sun to understand the business of guns, this surely had to be it.

The day before my meeting with the porn actress Stoya, I had secured a ticket for this monster of an event. That morning, I decided to walk there from my hotel, having booked into the cheapest room on the Strip, a slip of a room in the MGM Grand. But as this was

Vegas, I was taken in by an illusion. It was the magic of perspective that got me. The MGM is the largest hotel in the US, so it had the effect of making everything seem closer than it really was. An hour in, and I was still walking past drunks and lunging couples on their return from the night before. Then the mood shifted. My path was joined by stern-faced, bearded men. They wore khaki, black polo shirts and sculpted baseball caps. Wrap-around shades hid their eyes.

We fell into step, and I started chatting to one of them. His name was Jake. 'I survived the Gun Control Panic of 2013 and all I got was this T-shirt . . . and 20,000 rounds of .22 LR, 5,000 rounds of 5.56, 50 PMAGS, 10 stripped lowers, 3 reloading presses . . . and an angry spouse,' was written on the back of his T-shirt. He was from Idaho and ran an ammunition company. He had recently bought a reloading machine for just over a thousand dollars and had spent most of the previous year assembling the individual parts of bullets – hull, primer, powder and shot – then selling them on. It came out a third cheaper than factory-loaded ammo, and he had made $430,000 in sales. We were still ten minutes from the show, and you could already smell the money.

We walked into a mildly hallucinatory section of the Vegas Strip – a kitsch reincarnation of Venice, without the soul or the sewage. Passing concrete pink colonnaded bridges and barbershop-painted poles, Jake moved on ahead through a set of double doors. I followed. A casino lay between us and the convention centre, and the floor was lined with intensely coloured carpets, disruptive patterns designed to stimulate your eyes and keep you in a permanent state of wakefulness. Each minute you remained conscious was a minute more where a dollar could be stripped from you. There were no windows; the only stars Vegas wanted you to see were the Botoxed ones on stage.

The shrill sound of slot machines whirred in the close air. Some of the show's visitors peeled off to try their luck.[11] Even more joined the stream, and, like a shoal of salmon, the crowd surged and spilled down a staircase to the entrance.

There were 67,000 visitors here. In the first Shot Show in 1979, 5,600 people had turned up. But these vast, flowing lines consisted only of those who made a living out of guns. Casual visitors were

not allowed at this trade show. The tumult spoke of an industry of
a quarter of a million jobs, worth some $6 billion.[12] There are more
gun shops in the US than gas stations. At almost 130,000, there are
about ten times the number of federally licensed firearms dealers in
the US than McDonald's.[13] And the dealers are overwhelmingly
middle-aged, white and male; just looking at the legions here told
you that. There wasn't a black face in the whole crowd.[14]

I picked up an entry pass from an unsmiling man in a booth and
entered the show. As far as the eye could see, there were adverts and
banners, logos and pavilions all devoted to guns: 13 miles of aisles
filled with everything from small, family-run operations through to
international firearm consortiums. Here were gun producers whose
names conjured up cowboys and freedom fighters, despots and liber-
ators: Colt and Kalashnikov, Smith & Wesson, Heckler & Koch.

The show, prorated into huge halls, was divided like the steps of
my journey – hunters and sportsmen, police and military. A biker
pushed past me, swastikas tattooed onto his layered, fleshy neck. I
wondered if there was a section for criminals. I took a hard right
down the main drag, into the section for law enforcement.

The organisers had included this category twelve years earlier. Then
it covered 7,000 square feet; today it was twenty-four times that.
Dummies covered in SWAT team gear stood on all sides; enormous
banners displayed helmeted men with angry eyes. On each side were
logos and marketing creeds: 'Illumination tools that serve and protect',
read one for a torchlight company; 'For those who train with a higher
purpose', read another, and it felt like a muscular prayer.

I walked towards a booth selling SWAT equipment. 'Leave achievement and destruction in the path behind you,' declared its banner.
A mountain of a man, a gun trader from Oregon, was there, trying
on a bulletproof vest. His sides were exposed like slabs of butcher's
meat. He said that ever since Obama had spoken about looking at
the issue of gun violence following the massacre at Sandy Hook, a
sort of fever had gripped this country. Rumour had taken hold that
the government was coming after your pistols and your rifles.

'There was a frenzy of people coming through my doors. Obama was the best gun salesman in the US,' he said. 'I brokered deals for $500 guns that then sold for $2,000.'

An obese woman walked into me, forcing me back with her width. Her red T-shirt read 'I carry a gun 'cause a cop is too heavy'. The fat on her back caused the letters to bulge. I turned back, but the gun trader was already in another conversation. Business was brisk; cards and order forms passed between eager hands under company names like BlackHawk! and Warrior Systems.

I walked over to the Colt pavilion: a monument to American gun culture. Men formed a silent echelon around a slew of black matt guns. These were the ever-popular, ever-contentious, auto-loading, assault-style weapons – military rifles designed for civilian use. They had scientific names like the LE901-16S and AR15A4, and these men lifted them with focused skill. They knew they could stock their shops with guns like these.

So did the producers. In its 2011 annual report, Smith & Wesson said it saw a $489 million domestic, non-military market for these 'modern sporting rifles'. From 2007 to 2011, according to the Freedom Group – the world's largest gun conglomerate – US civilian rifle sales grew at 3 per cent a year, while those of assault weapons grew at a rate of 27 per cent.[15] It was no surprise, then, to learn that eleven of the top fifteen gun makers manufactured them.

The industry repeatedly says these semi-automatic weapons are for hunting and for target practice, but death – and the fear of death – stalks much of the civilian assault-weapon advertising. 'Survival means different things to different people,' said one of Colt's older ads. 'Take the shot of your lifetime!' the company urged.[16] Other companies did the same. 'Flat Out. Lights Out' was the tagline for one assault rifle. Another advert – brushed with crimson – was for a scope called 'Revenge'.[17]

Throughout the show death was a perpetual and unique selling point: killing here was marketed and promoted. A company called Gem-Tech produced silencers and said they were '62 Grains of quiet diplomacy'. One of their adverts had the catch-line: 'Our only trace is a body where the enemy once stood'. Another company showed the image of a sniper's rifle poking through a willowy grass verge. 'Do not underestimate the determination of a quiet man,' read its copy.

Alone, these images could be dismissed, but seeing row after row of them, I began to feel insanity creeping in. Profit here was infused with murderous intent. One company was selling sweatshirts with 'One Shot, One Kill' above a grinning skull. Others based their marketing on a quasi-Christian Templar iconography, like that of a crusader's skull next to the motto 'In Hoc Signo Vinces' ('In this sign we conquer'). The Holy Cross had been turned into a sniper's reticle. Another advert showed a shadowy figure, antlers sprouting from his forehead, a cup dripping with blood in his left hand – the stuff of nightmares. But the skulls these marketeers used were always bleached and intact. They were not the bullet-cracked skulls I had seen in San Pedro Sula, with their green and mottled flesh.

John Hollister, with his black T-shirt, shaved head and white goatee, looked the part in this alternative world. But he was not here buying. He worked for Advanced Armament Corporation based out of Georgia and sold silencers. 'We are a lifestyle,' he said, pointing to the logo of a skull above a pair of crossed AR15s fitted with silencers. 'We once had a deal where, if you got that image tattooed on you, you'd get a $1,000 credit from us. Two hundred and fifty people got tattoos in that first week.'

This was supposed to be a counter-culture, and yet even that had been appropriated, like so much here, by the logic of corporate capitalism. John said his office email signature read: 'A lifetime member of American Gun Culture'. And it was not just John who had this corporatised passion.

It was a profit-driven gun culture that was also at the heart of what Ed Strange, the manager of a company called Wicked Groups, did. Ed, like John, had a long goatee and wore lots of black. His arms were a patchwork of tattoos. He marketed his Michigan company as the 'Bad Boys' of gun grips. They customised the handles of pistols for people who wanted a mark of their individualism on their sidearm. But despite all the choice of customized art, he mainly sold two images – the US flag or the skull. His biggest customers were US soldiers spending their hazard pay or law enforcement officers pursuing private passions. And they were almost always white men. 'In this industry, I can't think of a single company that is African American,' he said.

This marketing of skulls, though, bothered me. It seemed as if death was abstract and saccharine. A bloody fashion accessory. And where there is death, there is always sex. 'Make Love Loudly, Make War Silently,' said the logo for a silencer manufacturer. The sexual marketing here was unreconstructed. Images of women in low-slung dresses holding cigarettes lined some of the booths. The scantily clad Hotshot Girls were out there in the mêlée, signing calendars. A Czech gun company had dressed up its pistol-toting models in bikinis and feathered wings and called them Guardian Angels. And the gun manufacturer Glock is rumoured to have taken this a step further by going to a local strip club in Atlanta and picking out the best-looking girl there to promote pistols at guns shows. And they did and she did.[18] I went over to their pavilion – filled with men taking steady aim with Glock's ever-popular pistol range – and asked for an interview. I wanted to ask them about this marketing tactic, as well as plenty of other things, but they just said they'd have to get back to me.

Patriotism was also present in swathes. The advertising dragged you back to the idea of liberty again and again. Freedom Munitions bore the tagline: 'Freedom starts here'. Satellite phone networks sold 'Freedom Plans'; $1,499 gun safes were called the 'Freedom Model'. The Stars and Stripes was appropriated in ways that demeaned not elevated it. Old Glory was used to sell everything from sniper optics to 'Kill 'em, Grill 'em' ammunition.

There was more than a whiff of Islamophobia as well. I had seen three target companies selling life-sized shooting mannequins dressed in keffiyehs and bishts. Mission First Tactical's stand had a poster on its glass partition of an American soldier in a desert, his head bowed. Behind him stood a woman in a burka. They were surrounded by phrases like 'I am a warrior' and 'I will always place the mission first'.

Of course, there were lots of outfits not bedecked with skulls and crosses, Mom-and-Pop concerns that had a much more considered approach to gun sales. But it was not moderation I was looking for. America's gun violence was immoderate, so my eye was drawn to its cultural peripheries, where I found a marketed obsession with death, faith and the flag.

But beneath the muscular Christian nationalism was a hard corporate reality, one without patriotic sentiment. No matter how much you wrapped the US flag around guns, you could not ignore the fact that three of the five top handgun firms at this show were not originally from America: Glock (Austria), Beretta (Italy), and Sig Sauer (Germany).[19]

This dichotomy was summed up by one image that repeatedly caught my eye. Framed photographs dotted around commemorated Mikhail Kalashnikov, the designer of the AK47. He had died a year before, and one photograph had a huge floral wreath of blood-red roses around it, placed upon a stand of two AK rifles. 'One of the greatest and most influential firearms designers of our time,' it read. It gave pause for thought. Here, in the heart of America, stood lavish memorials to a gun designer whose work had taken so many American lives, yet no one was protesting that fact. Six decades ago this would have got you dragged in front of the McCarthy hearings. Now it seemed the main person disturbed by the AK47's legacy had been Kalashnikov himself. He had, before his death, written to the head of the Russian Orthodox Church regretting his involvement in the twentieth century's greatest killing machine.[20] But here, it was business as usual.

It was intriguing. Was the allure of the gun so strong that the icon of communism and anti-America materialism – the AK47 – could be so easily embraced? To answer this question, I set up a meeting with Thomas McCrossin. He was the general director of the Russian Weapon Company and had just signed a deal with the company Kalashnikov Concern for the distribution rights of their rifles to the US. It gave McCrossin the chance to sell 200,000 Kalashnikovs every year from sea to shining sea, and I wanted to ask him about the image problem there might be in doing that.[21]

He greeted me with a suspicious look. A sombre middle-aged man in an ash-grey suit, he extended his hand and gave me that terrible handshake men sometimes do. The one when they force their hand to be the dominant one, above yours. I let him do it because it was a weak psychological trick, and it said more about him than me, and we sat down in a grey cubicle office that stood beside his company's rifle-lined stand.

Looking back at our meeting now, I still find it hard to decipher what he told me. The conversation was filled with leaden words such as 'private investors', 'consolidation' and 'exclusive distribution'. I asked how many sales he hoped to make. 'Leave that be,' he said with a push of his jaw. Instead, he talked about market levels returning to normalcy, artificial growth and 'sportarised' versions of rifles. Perhaps it was the windowless office, the blandness of his voice or my jet lag, but it was like skating on a clouded ice-rink in an ashen light. Imprecise, vague and elusive corporate-speak. The bloody realities of what these rifles had done, and could do, were reduced to a drab sales pitch. He then said he had another meeting and gave me his dominant handshake again, and I walked away, wiping my hand as if removing a stain.

Almost immediately, I stumbled across a far more modest stall. In many ways, it was the very thing I had been looking for – here was someone who might be able to tell me, without a PR officer getting in the way, about a world that was still hidden. One of exports and deals, shipments and arms trading – how guns get around the globe.

Unlike the other outfits at the show, this one was spartan: just a desk and a banner that read 'Hurricane Butterfly'. It was a small company run by a Chinese-American called Jason Wong. Trim, with a precisely ironed blue shirt and military-cut hair, Jason could have been an insurance dealer or an auditor. Instead he dealt in guns. His operation helped internationally trade firearms for those unwilling or unable to do it themselves.

'I get export licensing. I consolidate products. Then we ship,' he said, leaning back in the way men do when they are comfortable with what they do. 'I've sold about $5 million worth of guns. My industry is recession-proof – we sell to the entire world. This is a growth industry – it's not going away.'

In many ways he did the boring stuff. Things like facilitating shipments and securing end-user certificates, documents that certify the
buyer is the final owner of the guns, not just someone planning on
giving them to a rebel group or a terrorist cell. It was no surprise to
learn that Jason was once a lawyer. In this world you need someone
who has that focused patience in order to tell the difference between a
DSP-83 form and a BIS-711. He claimed he has a 97.8 per cent success
rate in DSP-5 export permit approvals. I had no idea what that meant.

But the things he sells are far from dull and orderly – and so he
has to contend with the perpetual scrutiny that he gets from the
State Department, from US Customs and Homeland Security, all of
them sniffing around. And he has to deal with the public's perception
of his job.

'When people hear what I do they get scared. "Have you seen
Lord of War," they ask? No, I haven't. I'm not interested. I only sell
to the good guys,' he said.

He exports to over twenty-five countries. 'There is a proscribed
countries list,' he said. 'We can't export to Syria, North Korea, Iran,
the Ivory Coast. So I sell to places like Guatemala, so its citizens can
buy guns with the hope it prevents a civil war in the future. This
might sound counter-productive, but if only the government has
guns, then how are you going to fight it?'

I asked him how he could be so sure the 'good guys' get the guns.
After all, guns exported by Hurricane Butterfly could get siphoned
off, fall into the hands of the wrong people. Surely this is what
happens when a butterfly flaps its wings – it causes a hurricane on
the other side of the world?

'Diversion is a naughty word. Does it happen? Yes. If it happens,
it is usually US government sanctioned. A lot of things go on that
people aren't aware of. It's not like the US government's going to
issue an export licence to the rebels in Syria. I know of colleagues
that were asked to legally supply arms to the rebels with US government sanction and foreign government investment. The US government is always interested in how the US government can get weapons
into foreign zones and they have to come to someone like me.'

'Someone like me' – the world of the middlemen summed up.

The corporate, articulate, besuited quietude. The neat world that lay between the pounding of a producer's weapon-making machines and the angry retort of those weapons used in anger. A world peopled by men who worked in clean, carpeted, well-lit offices, men with starched collars and cut fingernails who never raised their voices. As C. S. Lewis said, 'Hell is something like the bureaucracy of a police state or the office of a thoroughly nasty business concern.'

In the end Glock never got back in touch for that interview. It was a pity. I had questions to ask about what processes they went through when deciding what guns to sell and who to sell them to. I wanted to know more about how hellish bureaucratic police states got their weapons, and whether Glock could convince me that it was not a thoroughly nasty business concern.

I had my doubts.

On 1 September 2012 Corporal Dwayne Smart, a police officer in the Jamaican Police Force, killed Kayann Lamont. He was trying to arrest her for using 'indecent language' – a crime in Jamaica. She struggled, and eyewitnesses claimed that Smart shot her twice in the head. He then allegedly shot and injured her sister. Some were later to say he was in the middle of reloading his weapon to finish her off, when another officer intervened. It was to emerge that Kayann was eight months pregnant when she died. Corporal Dwayne Smart was charged with her murder.[22]

The thing that drew me to this ugly story was that Kayann was shot with a Glock service pistol, one most likely bought by that Caribbean state in 2010, possibly from a Glock subsidiary or agent.[23] So what, you may say. Gun companies make guns that kill people – it's self-evident. But Glock is a little different. The Austria-based company has become the leading sidearm of choice for police forces around the world, and the question that has to be asked is: does this company take into account who it really sells to?

The gun company has had resounding success in the US. Glock

itself boasts that up to 65 per cent of America's police departments put a Glock pistol 'between them and a problem'.[24] Given that almost 1,500 people were reported killed in police interactions in the US between 2013 and 2014, it's pretty certain that some of those died at the end of a Glock.[25] And not just in the US.

Glock pistols appear to have found their way, perhaps even without Glock's knowledge, into a number of countries with major issues surrounding their human rights records – places like Iraq,[26] Belarus,[27] Azerbaijan[28] and Israel.[29] It raised for me a specific concern that I wanted to ask Glock about: how much they, or their subsidiaries, took into consideration a nation's police human rights abuses before a contract was signed. Did they sell Glocks to the Jamaican police, for example, and – if so – did they know that force had been plagued by accusations of extrajudicial killings?

After all, Kayann Lamont was just one of 219 people killed in incidents involving the Jamaican security forces in 2012.[30] And Jamaican human rights groups have repeatedly been vocal about the need for more accountability over the use of firearms by members of their police and army.[31]

It is not just Jamaica's controversial police force that I wanted to talk about either.[32] In 2013, the Philippines National Police also reportedly bought nearly 60,000 Glock Generation 4 pistols.[33] In fact, another 14,000 Glocks were sold a year later, again to the Philippines National Police.[34] As I had seen, the police in the Philippines had a staggeringly poor human rights record.[35] Indeed, the police there are so riven with corruption that when the new guns arrived, the Philippines National Police chief reminded his officers that if any of them went out to pawn their guns, they would be charged with a criminal offence.[36] Were Glock entirely oblivious to these spectres of corruption and violence?

Perhaps. But that new Manila-bound consignment – predictably – was very quickly to make its visible, deadly mark. In August 2014, an off-duty policeman shot three men outside a bar in La Trinidad, Benguet, killing two of them. The gun used was the suspect's Glock service pistol.[37]

Of course, it is not just Glock who are clear that they are in compliance with strict export controls. Other European gun companies have been caught in the spotlight for selling guns to police forces with questionable human rights records.

In 2014 the German arms maker Sig Sauer landed in hot water with the authorities when almost 65,000 of their pistols were sold to the Colombian police via the US. Colombia at the time was on the German government's prohibited list of export countries.[38] In January 2014 the German police also seized documents from Sig Sauer that exposed the sale of seventy guns to Kazakhstan through questionable routes.[39] It was 'a total failure of controls', said the German Newspaper *Süddeutsche Zeitung*. 'Again and again, German weapons are found in conflict zones. And again and again the weapons manufacturers are totally surprised.'[40]

Another German gun manufacturer, Heckler & Koch, was also accused of supplying 9,500 G36 rifles to prohibited markets in Mexico between 2006 and 2009.[41] The German government had said H&K could sell arms to Mexico provided they did not end up in the states of Chihuahua, Jalisco, Chiapas and Guerrero. There was strong evidence that the police in these states had carried out 'disappearances' and extra-judicial executions.[42] But the Mexican Defence Ministry said they were unaware of the conditions attached to the rifles and so delivered the guns to the prohibited areas.[43]

The interesting thing is that Glock, Sig Sauer and Heckler & Koch were all working within legal parameters. In the European Union, weapons exports should, according to the law, take into account the 'respect for human rights in the country' they are shipped to. The trouble is that 'respect for human rights' is not clearly defined. And the gun companies' lawyers were adamant their clients adhered to the law.

But, as the Shot Show proved, getting direct answers to direct questions was hard. Harder than meeting an El Salvadoran assassin, harder than getting shot at. The situation is that, in general, detailed arms sales data is kept from the public eye, a lack of transparency that might even be bolstered by gun companies' often-close relationships with armies and police forces.[44] It means the effective scrutiny of arms transfers is sometimes impossible. We often know an arms sale has happened only after someone is killed by a gun from that trade.

In the European Union, where most of these gun companies are based, this lack of transparency is pervasive. Member states are obliged to submit data to the EU on arms transfers. Yet the quality and quantity of this data is far from perfect. In some years, France, Germany and the UK fail to make full submissions.[45] And Austria, home of Glock, has no published data on the UN arms register for six years in the last nine. Outside the EU it's as bad. US government information involving the export of 59,904 pistols from the American subsidiary of Glock to the Philippines was not released because the publication 'could cause competitive harm to the United States firm concerned'.[46]

If I could not get gun companies to respond to me, I figured I would just have to go to them. So the frustrating combination of silence and the opaque transference of arms led me to arrange a press pass to another type of gun show that intrigued me – the military arms fair.

'Welcome to Hell' said a flyer on the exhibition magazine. In the early June heat of a French summer, it seemed apt. There is something about the endless grey carpets and high spotlighting of convention centres that saps your time and then your soul. Row upon row of people hunched in their cubicles, staring at their laptops or at you – each partitioned by thin stall walls and each displaying an even thinner animosity for their competition. This was Eurosatory – the largest defence and security event in the world, with over fifty countries and 1,430 exhibitors selling the most advanced weapons systems man has dreamed up.

On the way to the convention, far out by Charles de Gaulle airport, I had taken a regional train. All the other trains were on strike, another French union disagreement. The carriage was filled, and it struck me:

Paris was not white. This car was rammed with Franco-Africans from Burkina Faso, Gabon, Guinea and beyond, many displaced by countless wars and conflicts. Some were self-consciously dressed in the dandy-jackets of the Congolese La Sap. Others were in colourful *pagnes*, eating cornmeal and banana fritters. Shoved in between them all were the white arms-sellers, clutching their briefcases to their chests.

By the time, though, that I reached the convention centre, a cathedral of chrome and glass and steel, the only black faces I could see were wearing the military uniforms of their African countries. None of them were selling arms. The white faces were doing that.

Perhaps I had expected an arms trade conference to have something that distinguished it: some tension, some geo-political animosity. But here the very reason that this exhibition existed – so that governments could rigorously attack or defend their sovereign interests – was hidden. The conference was broken up into national pavilions, and flags and banners stood high over the grey partitions; it felt more like the UN than a prelude to war. The vicious sting of international realities had been excised from this cavernous hall. The Russians and the Ukrainians were kept at a sensible distance. The Taiwanese were notable for their absence, the Chinese there in silent force. The Iranians and the North Koreans and the Sudanese were not there at all.

What was present, in abundance, was the stuff of war. This was expensive equipment. Global military expenditure was estimated to reach $1,747 billion in 2013, and duck-lines of military men with chequebook intentions strode through this space. They were filled with self-importance. Generals from all over the world in front, staff officers fretting behind. They eyed up the enormous trucks and tanks, the shoulder-held rockets and vehicle-launched missiles. They had a lot to choose from. There was an entire world of logistics: portable toilets, generators, kitchen units or quick-erect tents – the unexpected necessities of combat. These goods were sold with soulless straplines: 'Global solutions for local needs'; 'Tomorrow has arrived'. Here the marketing was stripped of that American ethos so overwhelming in Vegas. And without the Stars and Stripes, Liberty and God, the adverts felt bare; words like 'reliable', 'performance' and 'protective' spoke of staid procurement processes. Others were brutally frank: 'Dominating the Battlefield'; 'You can't choose your mission, but you can choose your equipment'; 'Making a risky job – safe!'

Some companies just struggled with how to market this stuff. Missile manufacturers handed out branded lip balm; body armour companies produced mini flack jackets for teddy bears. One just displayed a statue of the 'Predator' from the film franchise. Here the most diplomatic enemy to have was an alien – you never know who you might have to invade next.

Of course, national clichés and stereotypes lurked if you looked hard enough. The French army showed off their military rations – complete with terrine, fondue and cassoulet. The Austrians stopped for lunch at one o'clock, on the dot, and ate steaming-hot smooth sausages with glasses of beer.

National flavours aside, though, one logo summed up this whole exhibition: 'Connecting the Individual Soldier to the Battlefield Network'. The digital age had arrived in war-zones in a big way. You couldn't walk anywhere without seeing a drone or a surveillance camera or ways of killing someone without you having to be there. The remote warrior was all the rage. These weapons systems had PlayStation controllers, not joysticks, because this is what young twenty-first-century fighters are used to.

There was also a burgeoning innovation in firearms here, like the iPhone attached to the rifle that enabled commanders, sitting comfortably far away, to get streaming images of the battle; or acoustic detection devices that let soldiers identify the direction from which they were being shot at; or a digital sniper sight which ensured that you could only shoot when you were 'on target'.

But guns were still guns, and the new technological add-ons seemed just to highlight the fact that modern war was fast becoming a place where the pistol and the rifle felt increasingly redundant. Next to all the rockets and the drones, guns seemed diminished – old technology that actually meant you had to get involved in the grotesque business of combat. They certainly had a less-imposing presence here. Perhaps this was also because the contracts for small arms are far less financially significant than the cost of some battle tanks.

The show also told another story. Stripping away the Mom-and-Pop outfits of Vegas, it revealed how the deluge of guns emanates from a mere handful of the world's most powerful nations. What I had read during my research was strongly reflected here: the exports of no more than twenty countries account for 80 per cent of global gun trade.[47] It showed how, in the arms trade in general, the three biggest players were the US, China and Russia.[48] And, as I had already seen the US small-arms outfit up close, that left – for me – two pavilions to visit.

The Chinese pavilion was empty. A light from below cast a strange purple glow across the floor and made it look like a cheap Shanghai nightclub. Along a low shelf ran a line of armoured vehicle and supply truck models. Above them were glossy photographs of weapons systems with names like 'Sky Dragon'. This was Norinco's stand, one of four state gun manufacturers in China.[49] But there was a problem: ever since Tiananmen Square the French government had not let the Chinese sell their guns here. This was both a good thing and – selfishly for me – a disappointment.

China has long been a major gun exporter. In 2010 they sold about $89 million worth of guns worldwide.[50] At least forty-six nations had imported Chinese military guns in recent years, the largest proportion being in Africa.[51] But China is also repeatedly opaque when it comes to their government's gun sales. Published export data always shows Chinese gun exports as low for a state so often regarded as a major producer. This is because their guns are often sold at cut price as part of wider deals, sales are not recorded or weapons are just given away as gifts.

Despite this, some reports give a sense of what goes on behind China's often-impenetrable walls. In 2011, a UN report criticised them for relying too much on Sudanese government assurances that the weapons they supplied to Sudan would not – as they were – be transferred to Darfur.[52] Other documents found in Libya in 2011 showed that representatives of Chinese arms manufacturers met with Muammar Gaddafi's cronies. They had, allegedly without state approval, tried to sell the Libyans some $200 million worth of arms,[53] even though at the time Libya was subject to a UN arms embargo.[54] Then there was the delivery from China of guns for the Liberian special security services not reported to the UN Mission in Liberia. Instead the consignment was labelled 'spare parts and chemical products', and the UN in Monrovia was told it was a furniture delivery for President Johnson-Sirleaf's guesthouse.[55]

But there was no one in Norinco's pavilion able to answer anything. 'Please put your questions in an email,' was all I got. So I walked on.

The next pavilion was run by the Russians. In 2012 they made over $17.6 billion of international weapons and military equipment sales,[56] of which firearms made up about $157 million – mainly to the US, Indonesia, Germany, Brazil and Cyprus.[57] Unlike the Chinese, the Russians were there in force – tall, bearish men with hooded eyes and slender women dressed in high heels and fitted uniforms. The people from the weapons manufacturer Uralvagonzavod had built models of their armoured vehicles and were driving them up and down the hall. One was called the 'Terminator', and they had tied a Russian flag to its turret and were aiming it hard at visitors' ankles, making them step out of the way. At the time the Russians seemed on the verge of invading Ukraine.

Two Russian guns caught my eye. They were in a tank filled with water – a pistol and a semi-automatic – and below them was a sign: 'Underwater Small Arms System'. I had last heard of designs like these in the Leeds Armoury. Immediately, a woman called Angelina in a slim suit came over. She smiled delicately and asked what I was doing, and I tried to be charming and explained about the book. She was from the Central Research Institute for Precision Machine Building – or TsNIITochMash – a state-funded and Moscow-based guns research unit. They were a legacy from Soviet times, producing guns with precise and bureaucratic names such as the Gyurza SR-1M and the VSS Vintorez 'Special Sniper Rifle'. I wanted to know if I could visit their factory, and she left to consult.

Anna, the head of marketing, came over. Unlike Angelina, Anna was dour and suspicious, and her aggressive air was not helped by a lazy eye and a hunched back. She was dressed like the Bond villain Rosa Klebb.

'We do not likely allow foreign journalists to our concern,' she said. Then two men came over and crowded in close and pushed hard against me. Their fingers were stained with nicotine and their breath smelled. If this were a backstreet in Moscow it would not have been funny. But we were in Paris, so I wagged my finger and told them to calm down. They glowered.

I walked away and then found around the corner the stand for Kalashnikov Concern. I had had no luck understanding their US sales representative back in Vegas, so I asked if I could have an interview with someone here. Other hefty men started to crowd in on me, and I sighed. Russian diplomacy, I thought. Then, to my surprise, I was told yes, Andrei Kirisenko would give me a few minutes of his time.

Andrei was a professional shooter and the main counsel for the director of the group. He was enormous, over 6ft 4in. in his camouflaged deck shoes – bigger than the hired thugs at least. I wanted to know who they sold Kalashnikovs to. But he did not tell me. Instead I got more anodyne talk of 'modernisation' and 'product superiority'. So I asked about new technology. 'What is the next big transformation going to be in guns?'

'Guns are based on the principles of physics,' he said. 'We all want *Star Wars* guns but that will not happen until we can break the barriers set down by the limitations of physics.' It was a diplomatic response. The Russian attitude seemed to be – if it ain't broke, don't fix it. Not surprisingly – the AK design brought in profits by virtue of the fact that it was iconic. There were mountains of dollars to be made with that legacy.

So I asked him if he was concerned that Kalashnikovs were being used by forces the world over, including those with appalling human rights records.

'Of course, we understand that criminals use our weapons. But you can also get rid of bad guys with our weapons. Our weapon,' he said, perhaps getting carried away with his own voice, 'is a weapon of the world, it is a weapon of peace.' He rubbed his thick fingers together. 'Girls are weak, and so guys should protect them. I feel proud to be able to protect them with a gun.'

In the US guns seemed to be all about God, freedom and individualism. Here it seemed that power was what the gun really meant. Yet in both the US and here in Paris what was really evident was the constant search for profitability. God, power, flags and ego were just different ways to sell gunmetal.

Further away stood the Ukrainian stand, covered in strips of blue and yellow. A sign on it read 'We make Ukraine Strong'. If the Russians weren't going to talk to me straight, I thought, perhaps the Ukrainians might. I began to walk over, but then a bell sounded. Eurosatory was over for the day, and I had missed my chance.

But, in a sense it did not matter. Ukraine was a place I had already decided to go to on my journey, not least because, of all the countries in the world, it seemed to be the one place that was repeatedly implicated in the sale of arms to countries caught up in the tumult of war.

On 24 February 2001, the MV *Anastasia* was intercepted in Las Palmas in the Canary Islands. It was reported that Spanish officials had clambered on board and interrogated the captain. In so doing they reportedly found 636 tons of assault rifles, ammunition and other weapons in the dark hold.[58] The ship was duly impounded, at least until the Angolan government confirmed it was, indeed, their shipment, and the ship was allowed to continue its journey.[59] In 2001, Angola was in the middle of a civil war that was to claim over 500,000 civilian lives.[60]

In 2007, four days before Christmas, the MC *Beluga Endurance*, a ship flying the flag of Antigua and Barbuda, reported to be carrying 10,000 AKM assault rifles, as well as forty-two T-72 tanks, landed near the bleached beaches of Kenya's Mombasa. The load was allegedly headed for South Sudan.[61] A year before, hundreds had been killed in the South Sudanese town of Malakal in heavy fighting between northern forces and their former southern rebel enemies.[62]

Then on 6 March 2010, the BBC *Romania* was reported to have made its way to the mountain-rimmed port of Matadi in the

Democratic Republic of the Congo. It was carrying 10,000 Kalashnikov rifles. A few months later mass rapes were reported in North Kivu province. The UN envoy, a Swedish firebrand called Margot Wallström, blamed both the rebels and the DRC army.[63]

So far, so bad. Three ships scattered over time and destination, reportedly legally delivering guns to countries ravaged by violence, but there was one thing that they had in common. They had all begun their journeys in Ukraine.

Traces of these journeys, though, are not easy to find. The organisation that charts small-arms transfers – NISAT – shows virtually no records that these shipments ever left Ukraine. NISAT has no detailed data on exports from Ukraine in 2001. In 2007 there was nothing recorded having been sent to South Sudan – 40,000 machine-guns were listed as heading to Kenya and 1,000 rifles to Chad. Only in 2010 was NISAT able to list 10,000 sub-machine-guns and 3,000 rifles publicly as heading to the DRC.

It was purely through extensive data searches and cross-referencing that these shipments could be traced. But this was not new. The murkiness of exports from Ukraine has long been a source of concern to governments and arms trade observers alike. But an insight is sorely needed. Between 2006 and 2011 there were about $117 million worth of registered firearm exports from there, but this may be but a drop in the ocean of arms that have really left the ports of the Black Sea.[64]

Certainly, Ukraine has played a major role in the proliferation of small arms around the world. With the dissolution of the Warsaw Pact, Soviet military units left the lands they had occupied and headed home. As they headed for Moscow, countless guns were left unsecured and ripe for the picking. As much as 2.5 million tons of ammunition and as many as 7 million small arms and light weapons were left behind in at least 184 depots.[65] In Odessa, the imperial city of a million people, some 1,500 standard freight cars of ammunition were abandoned.

All in all, it worked out at about one hundred firearms being left for each Ukrainian soldier.[66] And so this newly liberated country began to offload its assets. Over the next six years, there was a reported

$11 million sales of small arms from Ukraine – most likely a fraction of the real amount shipped out.[67] In addition, when Ukraine gained its independence, its arms industry had to rethink who to supply next. Ukraine once accounted for about 30 per cent of the weaponry production of the Soviet Union, and there were about 750 defence industry enterprises with a staff of 1.5 million to feed.[68] This windfall inheritance and skilled workforce means today Ukraine is the fourth-largest arms exporter in the world, with about $1.3 billion sales.[69]

In turn, this industry created a highly efficient and secretive export system, one made up of opaque relationships between gun manufacturers, dealers, cargo companies, customs officials and off-shore financial services. This was why I had booked a ticket there.

Six kilometres south of the provincial city of Nikolaev, set between stony fallow fields and the thick, slow-flowing spread of the Volga River, is a little-known port. It lies fenced behind dense lines of barbed wire. A man-made forest further blocks your view of it. But as you get closer you catch glimpses of armoured bunkers and guards glaring at you from watchtowers and beyond. From the right location, you can even make out the outline of heavy loading cranes and thick earthen berms built to absorb the shock of an explosion. But if you venture too close, you come up against armed men with stony faces. There was one entrance, and there was no way I was getting in.

I had never expected to, really. This wasn't the place for prying eyes. But the one thing those shipments of small arms that had left Ukraine had in common was that all began their journeys here, in Oktyabrsk. And it was not just those three. This port had reportedly been the point of origin for repeated weapons shipments to over a dozen countries, many with a reputation for brutal repression: Sudan and Myanmar, Venezuela and the Democratic Republic of the Congo, Iran and Angola.

Today, it is estimated that up to forty weapon-filled vessels leave these private docks every year.[70] Even their website shows military cargo awaiting loading.[71]

At least Oktyabrsk was true to its past. For many years, it was a top-secret Russian naval installation. It was from here that Moscow sent missiles to Havana in 1962, triggering the Cuban nuclear missile crisis. And, despite being in Ukraine, Oktyabrsk is, as reported in the *Washington Post*, 'functionally controlled by Russia', run by a former Russian navy captain and owned by an oligarch with close ties to the Kremlin.[72]

This possible link to Russia was, to me, an intriguing one. After all, the Ukrainians were embroiled in a nasty standoff in the east with pro-Russian Crimeans. Why would there be so many arms shipments coming out of a Russian-run port in the heart of this country?[73]

I had emailed a few shipping companies to ask for an interview to find out more about this trade, but the same thing that happened with the gun companies in Vegas happened here. Nobody replied. But as many of these companies had headquarters in Odessa, two hours' drive to the west of Oktyabrsk, paying them a visit seemed my best chance of getting someone to explain these shipments of arms to me a little more. So that's where I headed.

The photographer's voice could just be heard above the roar of the fountain, and the bride shrieked as the wind picked up and sent a spray over her. Her hair was tight on her head and pulled back her features, making her look stern. She came over and asked me to get off the bench I was sitting on, so that she could get a better picture. Her bridesmaids, all dressed in the blue and yellow of the Ukrainian flag, stared at me. I wasn't going to ruin the photo opportunity, so I moved.

Starting my visit to Odessa with the sight of a wedding felt apt. After all, it has a strong reputation as a place to find love. My internet searches for the shipping companies that operate out of here ran constantly up against a tide of adverts for young brides and the promises of a lonely heart being filled.

But love was not the only relationship that flourished here. Everywhere you saw signs of the coming-together of business and trade. Above the airport's passport control were three adverts for freight service companies, one showing a lorry fitted with aircraft wings. Others highlighted offshore financial services. Speed and discretion were the key offerings here.

Once it was grain merchants that had made Odessa the fourth-richest city in the Russian Empire. Now its exports had diversified. Women and guns were new lures that hooked people here. I smiled at the posing bride and walked down to the docks. Small groups of Japanese tourists had left their cruise-ship and were being shown around. I passed them, listening to their guide explaining how Odessa was a city of immigrants – more European than Russian, of course – and I carried on down through a shaded park filled with broken concrete steps and on to the port.

The Black Sea spread into the glare of the morning sun, and across the bay a line of rust-metal ships were slowly being unloaded. Unlike in Oktyabrsk, you could see here cranes swinging endless containers through the warm sea air. The screech of tortured metal sounded, and other cranes stood unmoving, like mute robots, idling beside ships with glittering names such as the *Bosphorous Queen*. But this queen had long ago lost her regal shine and listed lightly against the dock, decaying in the sharp light.

There was nothing to see here that revealed Odessa's gun trade to me. I pushed back, up the Potemkin Steps, the formal entrance-way to the city, past heavy men encouraging tourists to pose with white-tailed eagles, and made my way to Number 10 Bunina Street.

At this address stood Odessa's most renowned shipping company, Kaalbye. It was a firm that prided themselves on shipping high-value military cargo. The US Navy's Military Sealift Command once hired them to transport mine-countermeasure vessels to Japan and a coastal security ship to Cyprus.[74] And their fleet certainly transported cargo from Oktyabrsk.[75]

I passed through Odessa's charming, decaying neo-classical streets, until I came to the address – the Maritime Business Centre – a ten-storey building of glass and secrets. Three men stood in the foyer, one of them with a pistol on his hip. I asked for Kaalbye shipping and they showed me to a small office on the side. There sat a young woman, dressed casually in jeans and a T-shirt. She gave me a bright smile as I walked in. I explained that I had emailed the company and no one had answered, so she told me to follow her and led me past the armed guard to a discreet lift at the end of the corridor.

Taking out a key, she pushed the button for the top floor, and the door closed. When it opened again, we were met by two kitsch statues: Poseidon and Mercury in gold gilt. They stood in front of two gaudy panels of painted glass – maps of the ancient seas – flanking heavy double doors that led on to a reception. Entering, it felt like stepping into Alice in Wonderland's study: seven doors branched off to hidden rooms, and in between them badly painted seascapes were hanging on wood-panelled walls. An early medieval galleon stood in a glass cabinet opposite a silver globe – an attempt at refined taste that fell short.

Behind a thick, marble-topped reception desk was a carefully coiffured woman staring at me strangely. She was dressed in a purple dress that would have looked in place at a cocktail party. I walked up and explained I was there hoping to see one of the directors. She looked pleased and told me that everyone was on holiday. She was sorry, there was nothing she could do.

My attempt to get an interview had failed, so I left a number and was escorted out. But no director called me. Instead, I got a phone call from their lawyer. His name was Andrew Friedman, with an 'expertise defending investigations involving . . . foreign corrupt practices . . . export controls and contractor corruption'.[76]

'I am representing Kaalbye shipping,' he said. 'I represent the company in litigation in the United States.'

'Right, so how can I help?' I was in the dark. Did they need a lawyer to give me an interview?

'Because there is pending litigation over what has been written about them – they are going to decline to comment. There was a research firm in the US that wrote a report . . .'

'In what context?' I asked, interrupting him. I was annoyed.

'In the context of arms shipments.'

'Where?' I said. I must have sounded like an asshole, but he was getting to me.

'It's online if you search,' he said.

The research firm he spoke about was called C4DS, a Washington-based investigative unit. C4DS had claimed Kaalbye was connected to an arms transfer to Angola in 2001; to an unknown shipment from Russian to Syria in 2012; and to a transfer of arms from Russia to Venezuela in 2012.[77] I wasn't sure what that had to do with me and told him as much. I said I wanted to interview Kaalbye about legal armament shipments – that was all. I had no evidence that they were doing anything wrong.

'They are going to decline to comment.'

'Why would you need a lawyer to call me up to tell me that?' I said.

'I am not in any way trying to intimidate you. I am doing it because my client asked me to call you to let you know . . . rather than just kind of silently ignore you.'

It was a courtesy call to tell me that I wasn't going to get an interview. And, oh, by the way, we are suing someone. It felt heavy-handed, but I guess when you start looking at the international shipments of weapons, you don't meet pushovers.

And that was it. My attempt to get some clarity on how guns are legally traded ended with a lawyer on the phone. Heaven knows what I was going to run up against when I started looking at the illegal trade.

13. THE SMUGGLERS

Catacombs and criminals in Ukraine – anarchy in Somalia, shopping in Mogadishu's market for AK47s – human rights abuse in Northern Ireland – how governments get involved in gun smuggling – the Second Amendment's long reach – the plea for no more US guns in Ciudad Juárez, Mexico

In early November 2013 the Greek coastguard intercepted a Sierra Leone–flagged cargo ship called the *Nour-M* near the Imia islets of the cobalt-blue waters of the eastern Aegean. Allegedly, there were 20,000 Kalashnikov assault rifles on board, along with 32 million rounds of ammunition. The ship had left Ukraine a few days before.[1]

According to the vessel's captain, Hüseyin Yilmaz, their final destination was the Libyan port of Tripoli. He said the *Nour-M*'s cargo had been purchased by their Ministry of Defence, and all was above board. But Greece's media reported differently.[2] They said that the Syrian port of Tartus was listed as the ship's final destination by marine traffic systems, and the captain had typed Syria into the navigation system, changing it to Libya only after his boat was challenged.[3] If this was true, then the ship was breaking an arms embargo.

Like so much in the world of smuggling, it will be hard ever to know the truth. The *Nour-M* was intriguingly to sink within thirty days of its seizure, battered into the depths by a storm off the port of Rhodes.[4] The Greek authorities have never clarified what happened to the 20,000 rifles.[5] But if, as the authorities and media suspected, the *Nour-M* was indeed smuggling arms out of the Black Sea, it would have been part of a long tradition of Ukraine's involvement in international illicit activities, one that was focused around the urbane streets of Odessa.

In the first half of the nineteenth century that elegant seaport grew in wealth and opulence, buoyed by the profits of a vibrant free trade zone set up by the Russian tsar. They were to call it the Porto Franco, and soon Odessa became a global centre, perhaps *the* global centre, for the trade in illicit goods. Porcelain came from China, flowered perfume from France, heady wine from Greece and, of course, muskets and rifles were there aplenty, because where contraband of illegal pleasures are found, so will there be guns – either as a weapon to defend the profits of the haul or just part of the haul itself.

Riches followed. Odessa's citizens had travelled there from Austria and France, Italy and Spain to create a better life for themselves. And if smuggling was the way to do that, well then, that's what they would do. They did it well. When the tsar considered withdrawing his financing of the Black Sea state, they sent 3,000 of Greece's finest oranges to Moscow to change his mind.[6] Each fruit was wrapped in a parchment inscribed with a list of all the benefits of the port to him. This bribe in fruit worked. Odessa kept its privileged status and evolved into a city that blinded the Russian imagination with its light architectural beauty and its dark tales of illicit luxuries. So the virtues of bribery and smuggling were here, visibly reinforced in every new building and in every clandestine delight.

It was a hidden character of this city that was pointed out to me by Valentyna Doycheva, a bob-fringed and elfin twenty-eight-year-old history graduate, who worked as one of the guides at Odessa's Contraband Museum. Showing me around the museum's modest exhibition floor, Valentyna described to me the catacombs of Odessa – the longest in the world – spreading like entrails deep beneath our feet. They once were used to store the city's surge of contraband. The museum itself was confined to five small-chambered rooms in the basement of a townhouse. Lined with glass cabinets filled with smuggled items, like clocks stuffed with cocaine or honey jars filled with melted dope, it was a noble institution seeking liberal truth. When criminal behaviour is so endemic, it takes courage to stand up and point out what was right and wrong. Even more so to dedicate a museum to it.

This is especially the case when you realise that today the culture of smuggling and organised crime still lies deep in the soul of this city. Since the fall of communism, Ukraine has emerged as the place to go for illicit goods, singled out as the epicentre of post-Soviet arms trafficking.[7] It is a quasi-criminal city where the government's involvement in smuggling is more than just looking the other way.

'Today,' she said, 'we get guns coming in from Russia – illegally, of course. And these guns are brought in by different criminal groups. There are a lot of them in Ukraine nowadays. It's a pity.'

Certainly the Ukrainian government has been culpable. In 1992 a commission concluded the nation's military stocks were worth $89 billion.[8] By 1998 $32 billion of it had been stolen and resold. As Andrew Feinstein wrote in his book *The Shadow World*, 'So explosive were the [commission's] findings that the investigation was suddenly closed down, seventeen volumes of its work vanished and its members were cowed into silence.'[9]

Behind this massive theft was a new breed of men – the so-called 'Merchants of Death'.[10] When the Cold War thawed, arms smugglers with names like Victor Bout and Leonid Minin swooped. Huge stocks of Ukrainian weaponry were bought up and sold to groups like the Revolutionary United Front in Sierra Leone and FARC forces in Colombia.[11] Men like Minin became major brokers of arms to Charles Taylor in Liberia, a country under an arms embargo. Minin did things like send 9 million rounds of ammunition and 13,500 AKM rifles to the capital Monrovia in two air-freight deliveries, listing them as headed for Burkino Faso.[12]

Today, though, things have changed. Experts have confided in me, in a way where even things that were not secret were phrased as being such, that the age of the Merchants of Death has ended. Instead, they said, new realities have created a different type of smuggler – often even more explicitly sanctioned by governments. One arms dealer said, 'I don't know if there could be another Viktor. He rose through the ranks at an opportune time. It was a sort of serendipity – the right time, right place. Viktor was there when the governments fell – he had connections. There is no Soviet Union today with vast quantities of surplus. The situation isn't right for that now.'

Today the global illicit trade in guns is just as prolific, but possibly more diverse. It happens at the point of production, where guns are stolen from the manufacturers, or – as in Pakistan – produced in extensive unregistered gun industries. It happens later – when guns are smuggled out of police and military arsenals, or seized when rival factions clash with each other.[13]

'Sometimes the goods are lost on the way,' said a Black Sea gun manufacturer I met. 'One of the things you have to look for is when small arms are taken by sea vessels. If there is a direct flight, then if it goes via sea it makes no sense. That's a tip-off something is going on.'

Then he told me a story. There was a time when a consignment of arms he had sold, in the belief the guns were headed to Jordan, ended up being earmarked for Libya. They would have been sent there, too, had the shipping company not inadvertently emailed his company a copy of the bill of lading. It claimed the container was filled with packets of soap powder, not 9mm pistols.

Of course, stemming such illegal flows of firearms is incredibly difficult. Unlike cluster munitions or landmines, guns have legitimate police, military and recreational uses. You can't just ban their manufacture and sale. This means it's even harder if a government is working in cahoots with the smugglers.

A few months before, I had met a man called Daniel Prins. He was chief of the UN's Conventional Arms Branch and headed up a department that oversaw attempts to prevent the trade of illicit weapons, without infringing upon their legal use and trade. In 2006 the UN had reported that a quarter of the $4 billion annual global gun trade was illegal.[14] He had his work cut out.

'I'll give you a hypothetical example,' he had said as we sat in a New York diner a stone's throw from the UN's grey-glass building on First Avenue. 'If you're an arms broker with a passport that's from Ukraine and you work from Cyprus, but your bank account is in the

Virgin Islands, and you broker a deal for arms that are made in the USA but are being shipped from Bosnia to Sudan . . . if you want to deal with this from a law enforcement perspective, where do you start? Whose regulations, whose laws do you need to follow? Is it the laws of Cyprus, where this person happens to be?'

He was a carefully spoken man, but frustration infused the little he did say to me, weighing his words down. 'The Russians tell us that they don't have a problem with arms brokers, because only one company is allowed to be an arms broker in Russia. But isn't it true that there are half a dozen, if not more, brokers around the world who do their work? And the answer that we get is: well, yeah, but that's not in Russia.'

Looking back, I wondered if he despaired in his New York office of all the traders in Ukraine who did Russia's dirty business.

'Globalisation means that with a cell phone and a computer it's easy to work from anywhere and to do your shipments and organise illegal shipments in a whole different way than it used to be, let's say, during the Cold War,' Daniel said. 'Nowadays, middlemen arms brokers don't need to be where the arms are.'

That was why places like Odessa were so attractive to so many smugglers. There, they could avoid controls. It was a place where the government would look the other way, at least if you filled their pockets when they were doing so.[15] In Odessa's tree-lined streets, criminals had the best of it. The weather was temperate. Women renowned for their beauty stalked the city's squares with lithe legs and feline eyes. There were nightclubs and restaurants to blow a bootlegger's profits. And, across this gilded and decaying city, there was easy access to supply chains, front companies and dubious handling agents prepared to do fraudulent paperwork. You could casually commit evil here without ever seeing the impact you were having.

Odessa had shown me what happens when governments allow their ports and businesses to become a logical extension of vast criminal networks and the corruption of officials permits unofficial trafficking. But, I thought, despite all of this, at least Odessa had the semblance of order. I had seen places where order had totally failed and where the smuggler had flourished: not because of the state, but because of the absence of one.

The only flight into Somalia from Kenya was an aid flight, and our TV crew were the only passengers. As we swung low over the flat spread of bush and shrubs that marked the Somali landscape, the emptiness and endless sand scudded beneath us. Just bush and the occasional dot of a herder with his cattle. Then our plane banked, and we headed sharply down to a line of sand and rocks traced far out on the outskirts of the ruptured capital of Mogadishu. It was the improvised international airport, without border guards or terminals, for the most lawless country on earth.

As we climbed down from the plane, six Somalis in battered pick-up trucks, each armed with machine-guns, met us. One leaned onto the ugly, heavy anti-aircraft weapon that had been bolted onto the top of his car. It was 39°C, but you felt the tension faster than you felt the heat.

I was part of a BBC team, fronted by the reporter Simon Reeve, making a series called *Holidays in the Danger Zone*. Given the number of guns here, the title was apt: weapons were in everyone's hands. About two-thirds of Somali men had at least one, and some estimates put the total national number as high as 750,000.[16]

Open-air gun markets were also common here. And we had come to Somalia in part to visit the biggest of them – a place where over 400 arms dealers sold smuggled AK47s to anyone who could afford one. Our film was about how a state survives when its government and entire infrastructure have collapsed. Somalia, wrecked by years of civil war, was the perfect example of ungoverned chaos.

Given the rise of Islamic militants and unruly warlords here in this blighted country, we feared kidnapping or worse, so we had hired guards. They chewed on sticks of khat, a local natural quasi-amphetamine, and fixed us with steady, hepatic eyes. We clambered up beside the machine-gunners and, exposed and burning under the unwavering sun, skidded through the dirt and dust to our fortified hotel. On the way they told us we were the only white people in the capital.

The streets quickly showed how the rule of law and those things we often take for granted – sewage removal and rubbish collection – had failed here. Foot-soldiers of the militia groups stood on crumbling street corners, young boys who had yet to buy razors but had long ago picked up semi-automatics. Everywhere you saw the violent marks of rivalries; I struggled to see what soul this city had worth fighting over.

There was no street electricity; abandoned houses stood in skeletal silence; potholes marked the roads. Outside our hotel was an upturned cow, distended in the heat. Someone had shot it a few days before, and the putrefaction touched your mouth.

The hotel had four guests. Me, Simon and our researcher – a feisty and huge-hearted Uzbek dissident and journalist, fluent in Russian and Arabic, called Shahida Tulaganova – and someone else: a Japanese government agent. I had sidled over to him and asked what he was here for. He told me he was a cartographer. I laughed and said nineteenth-century spies used to call themselves that. He never spoke to me again.

I was shown to my monk-like cell; cheap painted white walls and mildew marked the room. In it was a bed whose sheets had yet to dry. I lay upon the dampness and tried to ease that growing sense of panic that comes when you parachute into a place where, it feels, every other person wants to shoot you.

Later, towards dusk, we met in the hotel courtyard. High gates and walls rose on all sides, topped with shards of glass. There was a problem. Before we had set off from London, we had agreed a day rate of $50 for each of the guards, but the group wanted an extra $50 a day for their khat. At first I refused – I couldn't see a BBC licence fee payer being happy with us paying for the drugs of armed militia – but then I was taken aside. It was made clear that not paying would have consequences: our armed guards would no longer guarantee our safety. The emphasis was on the armed, and the threat was unequivocal.

Somalia in 2004 was not a safe place to be. If you were white and not a Muslim, it was more dangerous still. Within two years, two journalists I knew had been shot and killed there. Kate Peyton, a BBC reporter, went there just after us. She was shot outside the hotel where we were staying. Martin Adler, a Swedish cameraman, was gunned down a few months later in a crowded rally outside the national stadium. It wasn't the place to quibble over a few dollars.

I agreed a 'consultancy fee' and told them what they did with the money was their business, not the BBC's. The matter was settled, and we went inside for a meal of goat and spaghetti and arranged our visit to the smuggler's market.

The next day, as the call to prayer infused the city air, we set off, the guards making a small show of the khat in their mouths. Soon we came to a crossroads. Coming up the other road was a group of heavily armed gunmen – thugs from the enemy warlords. Our two convoys passed, and they immediately swung their guns onto us. The air was filled with a sharp cocking of rifles, and the occupants of the two cars began to scream at each other like dogs. Then one of our guards turned the anti-aircraft gun on them and – suddenly – the stand-off ended. We sped on our way. When it came to guns, here it was survival of the biggest.

Fear slowly eased back into fascination. Brightly painted trucks lined the road, horns blaring, fenders brilliantly painted in whirls of yellows and reds. We followed them, swerving past donkeys pulling wagons of rice and maize, until we came to an area where lines of ripe watermelons lay stretched out in the heat. On each side, pharmacies sold handfuls of coloured pills. Further along was the meat market, thick with flies. Men peeled off strips of crimson flesh with sharp knives, and thick white fat fell to the floor. There was no government here – this was the logic of trade at its most basic.

We got out of the car and walked over to a mound of green, vicious cacti. If you looked inside the thicket, you could see the outline of a US Black Hawk Helicopter. It was downed in 1993 in the infamous mission where Navy SEALS failed to capture the faction leader Mohamed Farrah Aidid. Instead, the Americans ran up against a barrage of Somali anger and gunfire. Beyond was a line of skinny, gazing boys, their homes the colour of burnt sienna, long ago marked by bullet holes. Beyond them spread Mogadishu's infamous Bakaara market.

Whatever you needed, this warren of huts and shadowy deals had it. Drugs? No problem. A passport? Mr Big Beard, a wizened Somali trader with luxuriously crimson-dyed hair, could make one for you for $40. An extra $10 would get you a diplomatic one. AK47s? How many, exactly?

We carried on into the shifting gloom, passing clackety-clack wooden shacks as we headed towards the gun market. We had been told we were not allowed to film, and by this stage I wasn't arguing. A few minutes before, the marked cocking of a semi-automatic by a lean and fierce guard had been directed at me as I filmed one of the many loan banks that skirted the market. And here there were plenty more guns to cock.

We passed one stall. Rows and rows of Chinese rifles lay propped up in the half shadows. Previously there had been Russian AK47s sold here, trickling out from the military arms caches of the old Soviet bloc. Then it was Ukrainians, Albanians and Romanians who had made their millions shipping semi-automatics to Africa. Now it was the turn of Chinese dealers, and good money was here to be made.

This week the price for an AK47 was about $860, up 40 per cent from five months before. And that was just the lower range of the AK47 variant: the 7.62mm. Those rifles that chambered the 5.45mm cartridge cost much more.

It was worth noting these things, because the price of weapons in gun markets can tell you a lot about the security situation of a country; rising price tags can be like a canary in the mine, foretelling the drums of war. Before coalition forces invaded Iraq in 2003, basic AK47s sold for as little as $80. Three years on, during the bloody tumult, they were ten times that.[17]

Gun prices also inform of recent horrors. Post-conflict areas are often flooded with cheap arms – FN FAL rifles sold for as little as $500 in Libya in February 2012, down from the thousands of dollars they cost at the height of the conflict a year before.[18] So exact can the formula between violence and the value of weaponry be, you can almost gauge the levels of violence in Syria by the price of bullets in neighbouring Lebanon.[19]

The costs of guns, though, are often subject to greater influences – including rumours, stock availability and simply whatever the trader can get away with. The UN has cited AK47s costing as little as $15, but I've never seen them sold that cheaply,[20] whereas a fully automatic Chinese Type 56 Kalashnikov can go for as much as $10,000.[21]

What was more certain than the price was the fact that the guns sold here would one day work their way far beyond these wooden racks. Guns sold in Somalia have wound up in Sierra Leone, been traded into Liberia and then used in killings in the Ivory Coast. The longevity of the gun means that you can never know where, or to whom, it will be trafficked – something governments don't think about nearly enough when they arm rebels or overthrow regimes.

This brief glimpse of an anarchic market also showed the face of a very specific type of trafficking – one that experts call the 'ant trade', where plentiful shipments of small numbers of weapons over time can lead to the accumulation of large and potent arsenals. Of course, I wanted to know more in that market: where these guns were bought in from, to whom were they sold. But it was too dangerous to linger, to ask questions. So we pushed on. Besides, we had Somali passports to buy.

What Somalia had shown me briefly, though, was the shadow outlines of a lawless world where smugglers work without governmental control. But what happened when it was too much governmental control that was the problem – when states smuggled arms? For me, the answer to that question was found far closer to home.

I had not been back to Northern Ireland since I was eight years old. A third of a century. I had remembered it to be, as childhood memories had marked upon me, a place in black and white: a monochrome-clouded land where the streets were lined with colourless, sullen men in ashen overcoats. But on that day I saw no clouds and no grey.

The brilliance of a rare sun shone on this land and, on the drive into Belfast, the fields stretched out in an emerald patchwork, the green so intense you felt it.

Grainne Teggart, a deeply conscientious woman from the charity Amnesty International, had picked me up from the airport. In the middle of my research for this book and years after I had come back from Somalia, she had invited me over from London to talk about investigating human rights at Belfast's largest community festival. As we drove through country lanes lined with thick-wrapped hay bales and dirty-white farmhouses, hedgerows and hawthorns stretching on either side, she spoke about the violent history of this troubled land.

Soon we saw the place-names: the Falls Road, the Shankill. We were in West Belfast; Union Jack bunting and Sinn Féin colours would have given the streets a festive feel, if it wasn't for the barbed-wire-rimmed walls and the infernal history of the place.

The sun was the hottest it would be all year, and white men with shaved heads rested, arms crossed over stained tank tops, on the gritty streets. Above them, Grainne pointed out the murals of para-militaries in black who clutched their equally black rifles. Punishment was often meted out by these men, and it was always hard and bitter. Kneecapping, a malicious punishment of getting shot in the leg, was once so common that a local surgeon was compelled to go to the highest echelons of the terrorist movement and explain how to kneecap without permanently maiming. Aimed more at sending a message than ruining a life, the punishments by the armed gangs got a lot less disabling after that.[22]

The talk at the festival went painlessly. At the end of it one woman, a campaigner and writer called Anne Cadwallader, put her hand up and asked me about the British government's arming of Unionist paramilitaries. I said I knew precious little about this, and she offered to share some information with me. As promised, an email soon pinged through.

What Anne outlined was stark. She had shown how members of the loyalist Ulster Volunteer Force had collaborated with officers in the Royal Ulster Constabulary (RUC) and the Ulster Defence

Regiment (UDR) to form a gang based in the so-called 'Mid-Ulster Murder Triangle'. She described how they were responsible for the deaths of over 120 people between 1972 and 1976, many with weapons taken from British army armouries.

In October 1972 one government document that she sent detailed how armed men had carried out a raid on a joint British army base in Lurgan in County Armagh. They escaped with eighty-five high-velocity rifles and twenty-one sub-machine-guns. An army intelligence report later recorded how the weapons went on to be used in a number of murders by the Ulster Defence Association (UDA) and that collusion by Northern Irish forces in the raid was 'highly probable'. In just eighteen months in the early 1970s the British army was to record seventeen incidents of weapon theft by loyalists where institutional collusion was suspected.[23]

It was not just the stealing of arms that fuelled the violence. The British government appears to have been actively involved in securing guns for terrorists, too.

In 2012 the Ministry of Defence and the Police Service of Northern Ireland were sued by the relatives of six men shot and killed by a loyalist gunman in a bar in Loughinisland, County Down, in June 1994.[24] The families accused the British authorities of having assisted – or, at best, just having looked the other way – as 300 guns and 30,000 rounds of ammunition, and more besides, were smuggled into Belfast in the late 1980s.

One of these, a Czechoslovakian SA Vz.58 assault rifle, was used in that County Down pub attack.[25] The same weapon had reportedly been used in 1993 in another attack on a van carrying Catholic painters to work in Belfast, in which one man, a father of five, died and five others were wounded.[26]

According to lawyers acting for the families of those massacred in The Heights Bar, the weapon cache was sold by Armscor, the gun sales corporation of apartheid-era South Africa.[27] A deal had been struck between the company and loyalist paramilitaries after a British agent, who infiltrated the UDA for the British spy agency MI5, went to South Africa in 1985. The agent had gone there specifically to line up contacts to buy guns, a trip funded, the MoD were later to admit, by the British taxpayer.[28] The deal was struck in June 1987 after $445,000 was stolen by Unionist terrorists from the Northern Bank in Portadown. That money was reportedly used to buy the weapons, and the guns duly arrived the following November at Belfast's docks, shipped in from Lebanon.[29]

The smuggled guns had consequences. After the shipment landed, killings by loyalists rose sharply. In the six years before the guns arrived, loyalists murdered about seventy people; in the six years that followed, 230 were gunned down, many of them innocent by-standers. Within weeks of the guns arriving, UDA gunman Michael Stone had shot and killed three men at an IRA funeral in Milltown cemetery in West Belfast.[30] Stone had been armed with a Browning 9mm semi-automatic pistol 'of the same type as those brought in from South Africa',[31] along with a .357 Magnum revolver.

The trouble with all of this, though, is that the evidence surrounding the pub shootings was heard under the terms of a controversial British secret justice bill. This allowed British government lawyers to keep their evidence secret, disclosing it only to a judge behind closed doors, where it cannot be examined or challenged by other lawyers. So we may suspect a great deal but – as with so much in this opaque world of gun smuggling – we may never know the truth.

Of course, the IRA smuggled guns and murdered people, too, even bringing in their arms in coffins.[32] But what interested me more was that the loyalist murders showed evidence of the British state's involvement in the smuggling of the guns. For me, this was a chink of light into the deeply concerning reality of the supply of government-funded arms to fuel insurgencies and terrorists.

The history of governments supplying arms is one, understandably, filled with intrigue and interference. In 2012, Saudi Arabia was reported to have financed the purchase, in shipments via Jordan, of 'thousands of rifles and hundreds of machine-guns' to the Free Syrian Army from a Croatian-controlled stockpile of ex-Yugoslav weapons.[33] Diplomatic cables released by Wikileaks showed claims of a Moscow-led strategy to use organised crime 'groups to do whatever the [government of Russia] cannot acceptably do as a government'.

This included gun running to the Kurds in an attempt 'to destabilise Turkey'.[34] And, of course, the US's long record of supplying rebel groups stretching from Syrian rebels[35] all the way to the Taliban is more than well recorded. At one point over 300,000 Afghan warriors carried weapons provided by the CIA.[36]

Then there are those weapons that are not purposefully – at least seemingly so – let loose from state control, rather ones lost and stolen. In July 2014 it was reported that 43 per cent of 747,000 weapons given to the Afghan National Army by the US Department of Defense could not be accounted for. These weapons, including rifles and machine-guns, were valued at some $270 million.[37] This massive loss was partly put down to 'missing serial numbers, inaccurate shipping and receiving dates, and duplicate records'. But it is likely that President Obama was not surprised to get this news. In a briefing from the head of the military force in Afghanistan, Marine Corps General John Allen, the president had been told that the major concern out there was not an ineffectual police force or an incompetent military; rather it was corruption. That, in Allen's words, was 'the existential, strategic threat to Afghanistan'.[38]

But it was not just corruption that fuelled this monumental cockup. The US, perhaps so loyal to the concept that the right to bear arms was the only way to govern, had given the Afghans far more weapons than they needed. They handed over 83,184 AK47s, even though, at the same time, they were asking the Afghan military to shift to NATO weapons like the M16. In total, over 110,000 weapons were given that were deemed 'surplus to requirement'.[39]

Of course, the US was then to deny any responsibility following this irresponsible largesse. 'It is the Afghan government's responsibility . . . to determine if they have weapons in excess of their needs,' Defense officials said in their report.[40] But the report also acknowledged the very real possibility that US guns could end up in the hands of the Taliban. Certainly ammunition magazines 'identical to those given to Afghan government forces by the US military' had been found on dead Taliban fighters.[41]

It was not just Afghanistan where this happened. In Iraq the Pentagon lost track of about 190,000 rifles and pistols they had given to Iraqi security forces.[42] As the Small Arms Survey concluded: 'Weapons were shipped via private arms brokers into a context where the human rights situation had been steadily deteriorating and where the likelihood of diversion was high due to poor oversight and generally weak stockpile security.'[43]

The reality, of course, is that some of these weapons, the sale of which had lined the pockets of US gun manufacturers, now line the arsenals of Islamic terror groups.[44] Firearms found being used by Islamic militants include American M16A4 assault rifles made by FN Manufacturing and Colt Defense, and XM15-E2S semi-automatic rifles made by Bushmaster Firearms International.[45]

It's clear the US's love for the gun has much deeper consequences that take it beyond a merely domestic issue about its citizens' rights to bear arms. In Iraq and Afghanistan this love moulded an ethos on how to rebuild a nation – as if through a sheer weight of weaponry you could stamp out dissent, quash radical opinion and sow the seeds of democracy.

I could not help but draw parallels between the US government's firearm-heavy response to crime, with their SWAT-team police culture, and their firearm-heavy response to nation building. The Second Amendment had shifted from a domestic issue into an international strategy. And, as I found out on a trip to Mexico just after I had attended Las Vegas's Shot Show, the Second Amendment was also to have even deeper consequences there.

'NO MORE WEAPONS' read the tall metal-lettered sign that stood over the Bridge of the Americas. The road beside it led two ways. One took you deep into the Mexican city of Ciudad Juárez, the other lifted you over the Rio Brava and away into the US city of El Paso. The message was not for the Mexicans heading north, it was for the Americans whose guns were trafficked here, deep down Mexico way.

The line of cars was heavy, as there was no toll here, unlike on other bridges, and it caused the cars waiting to cross passport control into the US to trail far back into the haze. Mexican men in blackened smocks cleaned car windows in silence. The Franklin mountains stretched into the American distance, framing a listless Stars and Stripes flag that was pushed by a stubborn desert wind from the south. To the side, two chanate birds, the *Quiscalus mexicanus*, sparkled in velvet blue and attacked each other viciously.

The sign's off-silver letters were made from crushed firearms seized by the Mexican authorities. When the billboard was unveiled in February 2012, the then Mexican president, Felipe Calderón, asked the 'dear people of the United States' to help end the 'terrible violence' in Mexico. There had been 120,000 homicides in Mexico between 2007 and 2012, and most had been with guns.[46] And many of these had been US guns.

'The best way to do this,' Calderón said, his voice lifting in the wind, 'is to stop the flow of automatic weapons.'[47]

You can see why he made the plea. Mexico has virtually no firearms manufacturing industry, they have very restrictive gun laws, and there is just one gun shop in the entire country.[48] Yet the numbers of US guns that end up in Mexico is breathtaking – about 253,000 are estimated to be smuggled in annually.[49] It's not hard to see where they come from. On the other side of the 1,951-mile border lie 6,700 licensed US gun shops.[50] And there's good money in this, too – one study found that 47 per cent of US firearms shops were dependent to some degree on Mexican demand.[51]

The outcome of this was summed up by a US Senate report that concluded about 70 per cent of guns in the hands of Mexican drug cartels came from the US.[52] The point was reinforced by some cartel leaders boasting that they buy all their guns from there.[53] Clearly firearms have increasingly become the drug lords' weapon of choice. In the 1990s guns were reportedly used in 20 per cent of all Mexican homicides; today they are implicated in over half of all murders there.[54] North of the US border, too, way up in Canada, the same applies: 50 per cent of the guns used in crime there were smuggled into the country.[55]

In Mexico, there is certainly enough evidence to prove cause and effect between North American gun sales and Mexican gun violence. When an American federal ban on semi-automatic weapons expired in 2004, Mexican gun deaths increased by 35 per cent in the Mexican counties adjacent to Arizona, New Mexico and Texas – all of which had lifted the ban. But the homicide rate stayed about the same in the Mexican counties south of California, where a state ban on semi-automatic weapons had remained in place.[56]

During the US federal assault weapons ban the number of arms crossing south every year was about 88,000; after the ending of the ban this amount had increased by 187 per cent. It has been estimated the lifting of firearm sale restrictions north of the border resulted in at least 2,684 additional homicides in Mexico.[57]

It's not just Mexico where American's lax gun laws have had an impact. The US government found that 76 per cent of firearms they traced in Costa Rica in 2013 were either manufactured in or imported into the US. It was 61 per cent in Belize.[58] And in Jamaica, American guns are said to be dropping into Kingston like mangoes off a tree.[59]

I looked the length of the spreading line of worn-down cars filing patiently into the US. The other road, the one coming down from the north, was empty – the officials here just waved the cars through into Mexico. The lack of checks on southbound traffic, combined with constant demand for firearms from drug cartels, must have created a perfect storm for smugglers. It begged the question: how did these guns get into smugglers' hands in the first place?

Court records give us a glimpse of one of the ways it works. In 2008, the American Range & Gun shop in Pembroke Park, Florida, made a hefty sale to a shadowy buyer. A few days before, the purchaser had told the dealer, Victor Needleman, that he didn't think he could pass a gun background check. The would-be buyer said he had been 'in some trouble' when he was younger. No problem, said Needleman, just buy it in another person's name – an illegal process called a 'straw purchase'.

The buyer, in fact, had never been in that kind of trouble – rather he was an informant working with the US Bureau of Alcohol, Tobacco, Firearms and Explosives (ATF). He made it clear to Needleman that he wanted to send the guns to Central America.[60]

Needleman said that wasn't a problem either – he had already sold guns into Guatemala. He was not bragging: several of his sales were later traced to shootouts between gangs in which a number of people were killed. Needleman even boasted about a customer who had bought twenty-five AK47s at a time.[61]

Soon afterwards, the informant returned with a 'friend' and put down the $2,120 deposit for fourteen semi-automatic pistols. His friend illegally filled out the paperwork. When they returned to pick up the guns, they ordered twenty more Glocks. Then Needleman was arrested. It was an open-and-shut case. Needleman was sent down for nearly six years.[62]

Legitimate gun dealers doing crooked deals like this, though, are just a small part of the problem. The issue is much more widespread. In 2012 an investigation found hundreds of civilian gun owners selling tens of thousands of firearms every year on the internet without any background checks.[63] Private sellers, meanwhile, make 'occasional sales' or sell from a 'personal collection' in gun shows all the time. The distinction might seem subjective, but the effect is quite significant. In the US federally licensed dealers are required to perform checks at gun shows, private sellers are not. And those concerned about this have estimated that as many as 40 per cent of gun transactions are conducted without such checks.[64] A US government report concluded that gun shows were the second-leading source of guns trafficked into the illegal market.[65]

Certainly there is plenty of anecdotal evidence to suggest that these loopholes should be of concern to the US public. There was Ali Boumelhem, a member of Hezbollah, who was imprisoned for trying to smuggle US guns back to Lebanon. He had been buying weapons at gun shows in Michigan.[66] Or Conor Claxton of the IRA, who had gone to South Florida gun shows to buy guns to smuggle back into Northern Ireland.[67] Even an Al Qaeda spokesman has remarked on the gun show loophole, encouraging American jihadists to 'go down to a gun show at the local convention center and come away with a fully automatic assault rifle, without a background check, and most likely without having to show an identification card. So what are you waiting for?'[68]

The concern of Americans on the Mexican border, though, was not about what was going south. Rather they were upset at what was heading north. I had spent some time with the Minutemen Project in south-eastern Arizona – a group of activists set up in 2005 with a mission to monitor the flow of illegal immigrants across the border. One of them had invited me on a flight low across the border, and as we skimmed across the arid shrub, you could almost feel their paranoia. These men, with names like Chuck and Jim, saw the wave of 'wetbacks' – economic migrants from Mexico and beyond – as genuine threats to their safety and liberty. But the occasional ragged figures of illegal migrants that we saw from the air just looked pitiful and furtive, not armed and dangerous.

The Minutemen, and others like them, had intractable and hardline views. In one video posted on YouTube, someone called Commander Chris Davis told militia members 'to go armed'. 'It is time that we start taking back our national sovereignty . . . How? You see an illegal. You point your gun right dead at him, right between the eyes, and say, "Get back across the border or you will be shot."'[69]

None of them saw the irony in all of this: that being killed by a US firearm south of the border was far more likely than being shot by an idiot with a rifle north of it. None made the connection that what the Latinos were fleeing from – with over 52,000 unaccompanied Mexican and Central American children seeking to cross this border between October 2013 and June 2014 alone – was the very gun violence that US guns had helped facilitate.[70] As Alec MacGillis wrote in the *New Republic*: 'The surge of migrants coming to the US from Central America is being fueled in part by the movement of guns heading in the other direction, from US dealerships doing brisk business with the help of porous guns laws and a powerful gun lobby.'[71]

And it was to this new group – the gun lobby – that I wanted to turn next. These were men who oiled the cogs of commerce for small arms, who spoke both for (and sometimes against) the gun's place in the world.

14. THE LOBBYISTS

Strange animals in Maputo, Mozambique meeting an anti-gun campaigner and ex-child soldier – sickness in New York as the Arms Trade Treaty comes into force – an American lobbyist and a power breakfast – the National Rifle Association's rhetoric examined – how the pro-gun lobby hijacked the horror of Sandy Hook, Connecticut – meeting zombies in Orange County, New York State – an insight into the American gun psyche at the London School of Economics

It was 25 June 2014 and a national holiday. The streets of Maputo, Mozambique's seaside capital, for the most part lay empty. The solid *fin-de-siècle* buildings were silent behind their elaborate wooden doors, save for the occasional sound of plates being put down for an early lunch. Men played checkers outside one of the humidity-streaked homes, their game caught in the shade of spreading trees.

There was an air of peace so different from this day in 1975, when the Portuguese had pulled out. Realising that colonialism had had its time, they left, destroying cars and pouring concrete down wells, and in so doing they planted the seeds of anarchy and violence that were to blossom into a bloody civil war less than a decade later.

Fearing the communist ideals that captured the imaginations of Mozambique's newly liberated leaders, Rhodesia and South Africa set out to destabilise their neighbour. They manufactured a guerilla movement, the Renamo, and packed it with mercenaries and disaffected Mozambicans. But the white South Africans and Rhodesians had no wish to govern. Their intent was destruction. Roads were blown to bits, landmines scattered in their millions, and the country was flooded with guns.[1] Atrocities were committed on a horrific scale: children and adults were tortured and murdered, populations were starved, and over a million people were killed.[2]

I walked past the low-level villas with their balconies and pistachio-green and lime-yellow walls and, passing three unarmed security guards asleep in the languid light, entered a courtyard framed by purple frangipani. Artists were painting to the low sound of soft rock. It was an arts centre, a cooperative, and I had come here to see a certain type of response to gun violence. What I sought was out the back: a room filled with fantastical creatures. One was a cruel, hunched insect, another a pointed, furious dog. But these were no ordinary animals – their bodies were made from the wooden stocks of rifles, their legs the stripped-down parts of an AK47. One had a long tail made from the coiled spring of a semi-automatic rifle. Another had holes pierced into a trigger guard, forming beady eyes.

They were the work of Makulo, a forty-seven-year-old Maputan. It was part of a charity-funded project where old arms left over from the civil war had been collected, and in return the charity had given the gun owners farmyard picks and hoes. 'Guns for ploughshares' they called it. And the artists here had turned the tortured gunmetal into hallucinatory sculptures. Shovels and sculptures – a form of gun control you could hold.

Makulo wore a paint-stained overall, a heavy woollen cap, despite the heat, and chains of shells around his neck. 'I like to make art,' he said, stumbling on the English words. 'It relaxes my mind. It makes me forget everything: war, the fighting.' His vision was the liberal lobbyist's dream: art used to defeat violence. 'If you destroy the gun to create something of beauty, it is one way to help stop violence.'

Others had done this, too. Mexican artists had turned guns into musical instruments,[3] making shovels with which to plant trees.[4] In New York, guns were made into bangles and cufflinks.[5] AK47s were turned into watches for the liberal elite, costing $195,000.[6] But, as I admired the sculptures, the peace here was coming undone. In central Mozambique Renamo troops had recently began killing, and the spectre of war was rising fast again.[7]

The next day I met Albino Forquilha. He was, like so many Mozambicans, smartly dressed. A cream-suited, kind-eyed forty-five-year-old, he was, as I was, here as part of a conference promoting the end of landmines. He was a lobbyist for their clearance in Mozambique, but landmines were just part of what he did. His main focus was guns, and he had spent his life working on the destruction of those left over from the war.

We sat down at a creaking table in a restaurant beside the conference hall. In the next room, diplomats from around the world, their country's nameplate before them, were listening on translator's headphones, charting the battle to rid the world of landmines and cluster munitions. Nobody here but us was talking about guns, or about what was unfolding in the north. For Albino, though, guns were what he knew. He had once been a child soldier, forced to join a rebel group when just twelve years old. He had had to kill.

'They were kids who had tried to escape. The commander called us all around and said that every one of them had to die,' he said, his voice soft. So he had shot seven of them in the head.

He had a low opinion of guns. Twenty years before, he had dedicated his life to clearing his country of them. A lobbyist seeking a country without guns, his work had been extensive – over the years his charity had collected almost a million of them. For a time he had been funded by the Japanese, Germans, Americans, Norwegians, Swiss and Swedes – and they had destroyed a mountain of steel muzzles and wooden stocks.

'Most of the guns were Russian. We had South African and American guns, but the most were the Russians.' So many that the Mozambicans even put the AK47 on their flag. 'We once arranged a huge explosion to destroy some of them – over 2,000 weapons in one go.'

His work was far from done. 'The numbers of weapons are still estimated at more than a million – there are still many weapons in Mozambique that need to be collected.' But he struggles to complete his life's work. Few countries are supporting him any more.

'Now we don't have the funding to restart the project. This is the problem,' he said. 'The money that is given to combat landmines is huge, much more than to destroy small arms.'

He was right. In 2012, just over 3,600 people were killed or injured by landmines and explosive remnants of war around the world – the lowest number recorded since the tracking of landmine victims began.[8] This was, in part, because $681 million was given in support of mine action that year.[9] This was why, in Maputo, the mood was buoyant. Financially supported political will was working – money had been found to tackle the scourge of landmines.

Nothing like it existed in the international attempt to stop small-arms deaths. I once applied to a European government for a grant to do a global count of mass shootings. It was a modest proposal, but my application was rejected. Of course, there are countries funding ways to address the hurt that guns bring, but far, far less money is spent on addressing the pain and suffering they cause than is made selling them.

Despite this, there is still a small group of people – lobbyists against violence – who have dedicated their lives to challenging the guns' ubiquity. Underfunded and all too aware of the enormity of their task, they labour on, seeking to get gun reform on the agenda. But for them it was often a Sisyphean task. The complexities of what guns can be – a hunter's tool, a policeman's power, a soldier's life – and the vastly differing opinions that exist surrounding them, make them the hardest of all weapons to regulate. Nuclear weapons – sure, you can see why people can envisage a world without them. A world without guns? That's a world without people.

Later, outside the Centro Cultural Franco-Moçambicano, a brightly coloured building in blues and oranges and vibrant pink, I saw another statue. This one was built from old European gunmetal collected by Albino's group. The sculpture was caught in the moment of death – shot in the chest. In the gathering dusk, it cast an ugly, shadowed shape. An obscene piston of an erect penis stood between his legs: hyper-masculinity in death.

The Institute was playing an outdoor film. I peered through the century-old iron railings: *20,000 Leagues Under the Sea*. A clipped English voice crackled over the warm twilight air – Captain Nemo.

'To be of benefit,' he said, like a warning, 'goodness must be constant, forever building. It must have strength!'

It sounded hopeless.

I had no strength. The sight of the damp, peeling paint in the upper corners of the room and the slow drip of the tap did not help. Nausea flooded over me, and the ache spreading across my bones felt like a bad blessing. On the twelfth floor of a slowly decaying hotel on First Avenue, my hotel was close enough to the United Nation's building to see it from the windows. But I lay in a foetal position in the dimly lit room and wrapped myself in sweat-soaked sheets, unable even to open the curtains. Malaria or flu, I had no idea what I had, but it sapped me of all will.

Three hundred metres away history was being made, and I did not care.

It was late March 2013, and the UN General Assembly was on the verge of agreeing a new accord: the Arms Trade Treaty (ATT). Ever since it was formed, the UN has always been trying to reduce the impact of weapons around the world. Its first resolution, the shock of Hiroshima still clinging to the conscience of the world like an atomic shroud, was about disarmament. Now, teams of delegates and officials, charity workers and lobbyists were engaged in furious last-minute pleas and deals. The wording of the Arms Treaty was crucial, and they had limited time to agree upon them.

It was an ambitious attempt to regulate the global arms trade. Groups such as Oxfam and Amnesty said the Arms Trade Treaty could reduce the trafficking of weapons to outlaw regimes and rebel groups involved in horrific atrocities against civilians.[10] It wasn't going to end armed violence tomorrow, they said, but it was a start. Behaviours like the ones I had glimpsed in the smuggling worlds of Ukraine and Mexico and Somalia needed to be changed over time, and this was one way to do it.

As it was, the treaty went to the wire, hampered partly by

denunciations from Iran, Syria and North Korea.[11] But on 2 April 2013 it was agreed.

This all sounds promising, but, however noble its ambition, the treaty's global impact on firearms is far from guaranteed. As one lobbyist said to me, 'We need to keep in mind the limitations of the ATT. The Treaty is not about banning transfers; it is about regulating them. And it is only about their transfer; it does not set rules about domestic control once they reach their destination . . . progress will take time.'

There is also the issue of getting key countries to agree to it. By early 2015, of the top fifteen gun-producing countries in the world only six had signed and ratified the treaty.[12]

The Arms Trade Treaty was not the first attempt to address the global scourge of guns, either. While that accord was about a wider array of weapons – from tanks to missiles – the UN had also seen attempts to specifically address the harm that guns cause. The most hopeful of these was the 2001 UN implementation of something called the Programme of Action.[13] This was designed to encourage governments to combat the problem of illegal guns. It tried to ensure stockpiles were not plundered and spare guns were destroyed; that systems for marking and tracing guns were put in place; that things like end-user certificates were regulated so as to control the smuggler's trade. All noble sentiments. But its progress has been slow. In 2012 four of the top fifteen small-arms producers – Austria, Belgium, China and North Korea – did not send a gun report to the UN.[14] And some ask if the relationship of states with the Programme of Action is like 'a loveless marriage, going through the motions with a fair amount of apathy, resignation, and lack of excitement or novelty'.[15]

Another endeavour, the UN Firearms Protocol, also set out to address the illicit trafficking and manufacture of guns.[16] But, once again, of the top fifteen gun-producing countries in the world, only four – Belgium, Brazil, Italy and Turkey – signed and ratified it.[17]

The issue, then, does not seem to be the quality of the treaties and agreements on offer. Rather it is the lack of support for them from those who truly matter. Of the top fifteen arms producers in the world, only Italy has engaged fully in all three of the most important gun-related treaties.[18] Perhaps the reason for this is that for many countries the idea of regulating the world's guns sends their governments into paroxysms of concern. After all, guns are used in plenty of legitimate roles. Nations have the right to defend themselves, and there is always the issue of the personal right of self-defence, as Asher John in Pakistan or Gayle Trotter in Washington had argued. Though the protection of these rights should be weighed up against the overwhelming evidence that show more guns mean more murders.[19]

There is even some truth that comes from the mouths of hawkish American commentators: 'The awful consequences of disarmament must not be ignored. Northern dough-faces agreed with the southern slavocracy that black people should not have guns; American progressives agreed with Stalin and Mao that only the government should have guns.'[20]

I have been blindfolded and led down humid paths to meet rebel groups fighting for their independence in the southern Philippines, or sat and spoken to the armed guards of refugee camps whose presence is the only thing that stops a massacre from happening. And in each place you are struck by that deep quandary – that there comes a time when a gun seems to be the only thing that prevents a human rights tragedy, as well as threatening to cause one.

The right to own guns for hunting has also haunted UN gun trade agreements. Concerns have been voiced about the rights of indigenous communities to uphold their traditions – from the Maoris of New Zealand to the Inuit of Canada – and has been a persistent point of concern.[21]

The thing, though, is this: these UN treaties did recognise legitimate rights such as self-defence and hunting. The international community could have their cake and eat it. So the question is – why had so many gun-producing countries not signed up to the numerous gun treaties? The answer to this lies, in part, to the existence of a very unique breed of men: the well-funded pro-gun lobby.

Tom waited until I had ordered before he said what he was having. My illness had long passed, and yet I hesitated over what to ask for. Either he wanted to see how much I would eat and have less than me, I figured, or he just was not hungry. It's hard to say when you dine with a pro-gun lobbyist. Power breakfasts are more about power than they are about breakfast, evidently.

In the mid 1990s a global coalition of forty-four pro-gun groups formed to match the disarmament coalitions. They called it the World Forum on Shooting Activities and claimed to represent over 100 million sports shooters around the world, including members of that behemothic American gun lobby group, the National Rifle Association.[22]

Tom Mason was the World Forum's lobbyist at the UN. He looked the part – like a man born from the fertile imaginings of Tennessee Williams. A dapper and smooth-jawed lawyer from Portland, Oregon, a state with the strongest of the strong hunting and shooting traditions. He ordered porridge ('too many calories in granola') and began talking about how the pro-gun lobby impacted the UN's Programme of Action – the main measure designed to combat the illicit traffic of arms.

'A lot of very, shall we say, "liberal" governments – I'm putting quotes around "liberal", you could use the term "leftist" governments – and some anti-gun forces tried to institute the Programme of Action . . . that effort was curtailed to a great extent by John Bolton.'

John Bolton. The prominent American neo-conservative who once served as the US ambassador to the UN and a man possibly best remembered for two things: the first was the whiteness and the bushiness of his moustache, the second that he was made ambassador to the UN even though he thought that 'there is no such thing as the United Nations. There is only the international community, which can only be led by the only remaining superpower, which is the United States.'[23]

He certainly brought with him an arrogance of ideology that comes with any superpower. As Tom said, the right to bear arms was there from the beginning: 'John Bolton said any Programme of Action could not affect civilian firearms.'

In his opening speech at that debate, Bolton laid down a series of what he called 'red lines'. These were issues the US would not accept in a final document. They included rejecting any attempt to impose restrictions on the legal trade of guns, any limits on the sale of guns and any text that made the Programme of Action legally binding. His 'red lines' were so bellicose they caused Camilo Reyes Rodriguez, the Programme of Action's conference president and the ambassador of Colombia, to roll his eyes. He was disappointed, the ambassador said in his closing speech, that 'due to the concerns of one state' the rest of the UN could not agree on controls over private ownership or on preventing the sales of guns to non-state groups.[24] Bolton had got his way.

What is he doing now, I wondered? Bolton chairs the National Rifle Association's international affairs subcommittee. Of course he does.

From that august position Bolton has tried to trip up negotiations on the Arms Trade Treaty. He said, for instance, that the treaty would constrain the freedoms of countries that recognise gun rights, that it would 'specifically, and most importantly, constrain the United States'.[25] It caused Amnesty International to specifically ask the NRA to 'drop its campaign of distortions'.[26] But the damage was done: distortion was pervasive, rumours circulated.

These whispers suggested the treaty was set up by the UN to put a stake in the heart of the Second Amendment;[27] that it would lead to the US government creating a registry of gun owners;[28] that it was to be signed into law while Congress was not in session.[29] None of these was true. Even the American Bar Association's Center for Human Rights said it was 'unlikely the proposed treaty would compromise Second Amendment rights'. If it did, 'the treaty itself would be void'.[30]

Perhaps people saw through the lies. The NRA, for once, was unsuccessful at getting what it wanted. The Arms Trade Treaty was passed with an overwhelming majority, and the US itself signed up to it in September 2013.[31] But – and this was a big but – it had yet to be ratified by the US Senate. The *Wall Street Journal* said at the time: 'If the NRA loses . . . in New York, the organization would probably shift its focus to the Senate to prevent ratification of the pact.'[32] It was there, in the Senate, that a two-thirds majority was required for the treaty to be ratified. Certainly the NRA's Washington lobbyists had already begun their work.

Earlier that year, at three o'clock on a Saturday morning in March, a non-binding amendment was passed in the Senate with a 53–46 vote opposing the Arms Trade Treaty.[33] The largest gun producer in the world showed its colours in a pre-dawn vote: they are unlikely ever to ratify the treaty.

Russia was to do likewise. And when you consider that between 2008 and 2011 the US and Russia made almost 70 per cent of all arms transfer agreements to the developing world, this is a concern.[34]

I asked Tom about Bolton. 'So, was he coming from a domestic perspective – from the US Second Amendment?'

'To a great extent, yes,' said Tom. They call that lobby understatement.

Again I saw the Second Amendment having deep consequences far beyond the US's borders – not just in the smuggling of arms into Mexico and Central America, but also in the hamstringing of international treaties and its impact on debates on how to limit the harm caused by guns around the world. Because if the world's biggest gun producer isn't playing ball, you've got a problem.

This was not just by refusing to sign treaties. The NRA has a history of actively getting involved in pro-gun lobbying outside the US. It gave support to those in Ottawa seeking the closure of Canada's gun registry.[35] It gave seminars on public safety – one called 'Refuse to Be a Victim' – in Costa Rica and Trinidad and Tobago.[36] And when Brazil, a country with more gun deaths than any other country not at war,[37] tried to hold a referendum on a nationwide gun ban the NRA waded in. Lobbyists were sent down. 'Emphasize rights, not weapons,' was their mantra. So the Brazilian gun lobby began running adverts saying that if the government takes away your right to own a weapon, they could take away other liberties from you, too.[38] Before the campaign, over 70 per cent of Brazilians said they supported the gun ban. By the end of it, 64 per cent of Brazilians voted against it.[39]

Even Albino in Mozambique had been affected. 'Once a member of the National Rifle Association came to my office in 2001,' he said. 'He came not to destroy the guns, but to ask if we could give them to the NRA.' It staggered me that someone from the NRA would trek across the world to a country freshly suffering from a million dead and still offer to buy up their remaining guns.

So I asked Tom how he could object to guns being sold to governments with histories of human rights abuse. But Tom was as slippery as he was charming.

'It's hard to talk about a fundamental objection to human rights abuses when nobody really has defined what that human rights abuse is,' he said. 'It's a real possibility that some leftist or some anti-gun organisation could go to a government of a country that manufactures and exports firearms and say, "We are objecting to your exports of firearms to the United States because there are human rights abuses in the United States with firearms'."

One thing was clear. That to understand how the global war on guns has been largely hamstrung, I had to understand how the Second Amendment and guns are lobbied for in the US. And the greatest lobbyists there – certainly the ones with the most influence – were the National Rifle Association.

In 2014, at the NRA's 143rd annual meeting, the executive vice president, Wayne LaPierre, took to the podium. He looked smart in a striped tie and sombre suit, but the warning he gave was filled with fire and brimstone.

Some of the things he talked about were noble and good. He challenged workplace bullying and praised small acts of kindness. He said there were people in America who would walk past an abandoned child in the streets. 'Not me,' I thought. And not members of the NRA, said LaPierre.

It made you think the members of the NRA must be good people. After all, as LaPierre said, they get involved in Little League, go to church, obey the law. They don't ignore orphaned waifs. So what's not to like? He said, 'We are the good guys!' a dozen times.[40] And, according to a 2012 Gallup poll, 54 per cent of Americans also thought the NRA were the good guys.[41]

It's hard to disagree. In my travels around America, I had met plenty of NRA members who were good, solid folk – sometimes angry, sometimes suspicious, often scared, but overwhelmingly salt-of-the-earth people and charming and funny. They had a view of guns neatly summed up in a newspaper comment piece: 'When owning a gun is not about ludicrous macho fantasy, it is mostly seen as a matter of personal safety, like the airbag in the new Ford pick-up.'[42]

The audience was filled with men in check shirts with thick-set arms and big moustaches. They nodded and occasionally applauded. LaPierre was preaching to the choir here. But it was not the people who made up the NRA that interested me. Rather it was the agendas and methods employed by their leaders and lobbyists that fascinated. Even more so when they talked about things like personal safety.

At the Leadership Forum, LaPierre's tone got darker.[43] He said Americans were buying more guns and ammo than ever before. Not to cause trouble. No. 'We already know we are in trouble.' His words grew ominous.

'There are terrorists and home invaders and drug cartels and car-jackers and knockout gamers and rapers, haters, campus killers, airport killers, shopping-mall killers, road-rage killers and killers who scheme to destroy our country with massive storms of violence against our power grids, or vicious waves of chemicals or disease that could collapse the society that sustains us all . . . We are on our own . . . The life or death truth is that when you're on your own, the surest way to stop a bad guy with a gun is a good guy with a gun!'[44]

It was an apocalyptic vision. One in which he even warned he would be ridiculed for 'feverish fear mongering'. This was a clever ploy. If you pre-empt criticism as inevitable, you help reduce the power of that criticism. But his words confused me. LaPierre's rhetoric seemed to run counter to the NRA's mantra that more guns equals less crime. They themselves ran headlines that read: 'Gun ownership at all-time high, nation's murder rate at nearly all-time low'.[45] They couldn't have it both ways, could they? At the same time LaPierre was summoning up the nightmare of pervasive and increasing threat to life, America had never been more heavily armed. So why was he saying this stuff?

His speech was clearly not about logic and joined-up argument. It was about emotion. As Ana Marie Cox wrote in the *Guardian*, his underlying message was: 'Give the NRA money. Give us money so we can create the legal environment that allows gun manufacturers to make more money so that they can give us more money.'[46]

The NRA is certainly not a poor organisation. Nearly half of the NRA's funding comes from the dues paid by its 5 million members.[47] But its big cheques come from the gun industry. Campaigners claim that, since 2005, 'corporate partners' have donated as much as $60 million to the NRA.[48] Included in this figure are eight gun companies who have 'given gifts of cash totaling $1,000,000 or more'.[49] And you are struck by facts such as the NRA gives each gun industry CEO their very own golden jacket if they hand over a million bucks.[50]

It is a virtuous circle – or a vicious one, depending on how you look at it – where profits from purchased guns are used to fund lobbying that help secure your rights to purchase more guns.

And a good deal of lobbying goes on: in 2013 the NRA spent $3.4 million on that dark art.[51] The question is, of course, who is pulling the strings here – the manufacturers or the lobbyists? I asked Josh Horowitz, the executive director of the Coalition to Stop Gun Violence about this, and his response was direct: 'People always say, "Oh, the industry controls the NRA . . . they give them a lot of money." They do give them a lot of money, but it's almost like extortion money: "You will give us this money", or "You will not innovate for safety . . . to put better trigger locks on. Because if you do, we're going to boycott you."'

He had a point. During the Clinton administration, Smith & Wesson committed to help prevent gun sales into the illegal market.[52] That decision almost put the company out of business.[53] The NRA instigated a boycott, and Smith & Wesson ended up losing 40 per cent of its sales.[54]

Putting it simply, the NRA has the US gun industry by the balls.

No other gun lobby in the world has this sort of influence. When Nigel Farage of the British UK Independence Party called for handguns to be legalised and licensed in the UK he was met with derision.[55] When Germany introduced a firearm registry, the gun lobby response was muted. 'The German minister of interior promised to guarantee a very high level of security of the data, so for us it's not a problem,' said Frank Goepper of Forum Waffenrecht, one of the main gun rights groups there.[56] By and large gun lobbyists elsewhere are niche groups. In the US, the NRA is not only the main voice for gun owners, but it has a deep and pervasive influence on America's gun culture. This influence was touched on by LaPierre in his speech.

He had two messages that caught my attention. First, he criticised the media for deceiving the American public. Then, he said that US laws were being used to the advantage of political elites.

On the issue of left-leaning hacks he grew invigorated. You knew the media was lying, he shouted, because they still call themselves journalists. It was an intriguing finger to point. After all, the NRA are masters of the art of inflammatory disinformation. They talk about the 'spillover' of border violence and homicidal immigrants – without acknowledging the harm that US guns cause south of the border. They see the perceived threat of terrorism as underpinning the need for American citizens to bear arms,[57] even though in 2013 only sixteen US citizens were killed worldwide from terrorism.[58] The same year there were 32,351 firearm deaths in the US.[59]

Having said this: what do you expect? The NRA are lobbyists. Of course they exaggerate, just as liberal lobbyists will portray an opposing view. It's the nature of the beast. But there was one thing that, when it came to the media, I did find pernicious, and that was the NRA's stifling of legitimate, independent research.

In the 1990s there were serious attempts by the US government to look at the impact of gun homicides. But the NRA's position, then as now, was that this was an 'abuse of taxpayer funds for anti-gun political propaganda under the guise of "research"'.[60] The tipping point seemed to come when Art Kellermann, a medical researcher, found that guns kept at home were significantly more likely to be used to kill a family member than to be used in self-defence.[61] The NRA response to this was to crush any dissenting research.

They focused their attention on Centers for Disease Control and Prevention (CDC) funded gun research.[62] Lobbied by the NRA, Representative Jay Dickey, a Republican from Arkansas, pushed through an amendment that said: 'None of the funds made available for injury prevention and control at the CDC may be used to advocate or promote gun control.'[63] The CDC's funding was slashed by $2.6 million, the same amount they had spent on gun research the year before.[64]

These effects are long-reaching. A pioneer in the field of injury epidemiology, Dr Garen Wintemute, who had his CDC financing cut in 1996, told the *New York Times*: 'The National Rifle Association and its allies in Congress have largely succeeded in choking off the development of evidence upon which . . . policy could be based.'[65]

The NRA have choked other things, too. They were behind a 2003 governmental edict that said the ATF, the body that regulates firearms, could not give out to researchers any data about the tracing of guns involved in crimes. They were also there when the FBI was told they had to destroy records within twenty-four hours after Americans passed a gun background check.[66]

But it was not just an information war that the NRA was fighting. LaPierre also used the podium to lambast Washington's abuse of political power. He talked about the law in the US being 'selectively enforced', clearly with an eye on Obama's desire to impose some form of gun control to reduce events such as Sandy Hook. This accusation concerned me.

After all, the NRA has major political clout in its own right. Eight US presidents have been lifetime members.[67] All but three of the forty-five American senators who torpedoed gun control measures in Congress in 2013 had been paid by gun lobbyists.[68] And, according to the Washington-based Center for Public Integrity, the NRA and the firearms industry have pushed over $80 million into political races since 2000.[69]

The impact of this lobby can often be seen in the absence of, rather than the presence of, politics. The US firearms industry is remarkably unregulated. The Consumer Product Safety Commission – a US government agency that protects consumers from 'unreasonable risks of injury or death' – does not have the authority to regulate guns, while the 2005 Protection of Lawful Commerce in Arms Act means that gun manufacturers are shielded from many product-liability lawsuits. So toy guns may have mountains of regulation to reduce the risk of them causing fatal accidents,[70] but real guns have no federal safety standards at all – even though American children are sixteen times more likely to be unintentionally shot and killed than children in other high-income countries. It was a cruel conundrum that was down to the singular power of the firearms lobby.

It goes further. Cigarette manufacturers are not allowed to market directly to children, but gun manufacturers very much are. Keystone Sporting Arms' Crickett rifle is sold as 'my first rifle', using a cartoon cricket as its logo. The NRA offers shooting camps for kids and publishes a magazine called *NRA Family InSights*, with a special section for under-eights. All of this ignores the fact that gun injuries send about twenty American children to hospital every day;[71] that in the US guns kill twice as many children and young people as cancer, five times as many as heart disease and fifteen times as many as infection.[72] And it was in this painful conjunction of child deaths and political power that I found the NRA's lobbying influence truly bewildering and painful.

The trees bore the memories of what had happened here. Other towns in this part of Connecticut wore the pleasures of Halloween happily. Ghouls and goblins were bought in bulk at Walmart, and the cartoon horrors of the American gothic imagination were given full reign. From trees hung grinning skulls or glowing pumpkins.

But not here.

This year the decorations were tamed in Sandy Hook – ribbons replaced skeletons; there were lights instead of dripping blood – because less than a year before, Adam Lanza had killed twenty-six children and teachers at an elementary school just up the road.

Driving into Sandy Hook felt like entering a forbidden territory. I had wanted to go there and walk the streets, browse shops and speak to locals about what had gone on here in the winter days of 2012. But I could not leave the car. I felt like an intruder, and a cheap one at that. Perhaps I had been to too many of these silent streets by now. The memories of Finland and Norway still remained. What had started as journalism had shifted into something darker, and I felt I had no place here.

I passed the school, and there was a sign outside. It was a construction site now, and no unauthorised vehicles were permitted to enter – they had pulled it down. I carried on driving until I had passed the outer reaches of the town. There I parked and walked into a local coffee shop.

The Starbucks of Newtown is, for the most part, unremarkable. A poster of Edward Hopper's *Night Hawks* hangs on one wall, a brightly coloured collection tin for the 'Faith Food Pantry' stands to the side. I walked up to the manager, a bald and smiling man. He quickly realised why I was there and immediately said he was sorry, he couldn't talk; he would be uncomfortable speaking to me. He echoed my own feelings there and then.

I had come to this particular coffee shop because here, less than a year after the shootings, over two dozen gun rights supporters had converged to voice their appreciation of Starbucks' policy not to ban firearms in its stores. Some came wearing camouflage and packing pistols. They were to be disappointed. It was shut – a sign read: 'Out of respect for Newtown and everything our community has been through, we have decided to close our store early today.'

Their actions, though, became national news. Locals were outraged. Elsewhere Starbucks had drawn plenty of criticism from anti-gun groups for allowing people to openly carry guns in its stores in states that allowed it. But for pro-gun owners to come here – here, where the soil still lay freshly turned on graves too small to look at – to make a political point? That was unfathomable.

I struggled at times like these to find any common ground with the pro-gun lobby. And it was not the only thing that gun lobbyists did that was offensive to some. Two groups even scheduled an event called 'Guns Save Lives Day' for 14 December 2013 – the first anniversary of the Sandy Hook massacre.[73] Another gave away guns to residents in Orlando, Florida, just 20 miles away from where neighbourhood watcher George Zimmerman controversially had shot and killed Trayvon Martin, an unarmed black teenager.

There might be outrage in such actions, but spotlighted shootings in the US follow a depressingly similar pattern. Demands are sounded for action to address the availability of weapons. The debate runs quickly up against a wall of opposition on the right to bear arms. And the pro-gun lobby wins the day.

Elsewhere in the world mass shootings have provoked governments to introduce ways to prevent future mass shootings. After the Hungerford and Dunblane massacres, in England and Scotland, respectively, the British government brought in stricter gun controls.[74] When fourteen were killed in Aramoana, New Zealand, lifetime gun licences were replaced by ten-year ones. The massacre of sixteen in Erfurt in Germany in 2002 led to the screening of buyers under the age of twenty-five for psychological concerns.[75] And in the mid 1990s in Australia, the Port Arthur massacre led the Conservative government to ban automatic and semi-automatic weapons as well as initiate a nationwide gun buyback scheme.[76]

These laws worked. Firearm homicides in Australia dropped 59 per cent between 1995 and 2006. In the eighteen years before the 1996 laws, there were thirteen gun massacres resulting in 102 deaths. Since the introduction of their laws there have been no massacres.[77] In 2008 to 2009, there were thirty-nine fatalities from crimes involving firearms in England and Wales, with a population about one-sixth the size of America's. In the US, there were about 12,000 gun-related homicides in 2008.

The US, though, is different. It is the only country in the world where, following a mass shooting, the nation has responded with loosening, not tightening, gun laws. After twenty-one people were killed in a mass shooting in Texas in 1991, the state pushed through a law permitting the carrying of concealed weapons. Other states followed.

Even the massacre of Sandy Hook saw a call for more, not fewer guns. Twenty-seven American states since that fatal day have passed ninety-three laws expanding gun rights, including measures that let people carry concealed weapons in churches.[78] Some schools even allowed their teachers to go armed.[79] So fearful were many US firearm owners about gun control that they bought up millions of rounds of ammunition. In fact, they bought so much that the global supply was impacted – even Australian ammunition stocks ran short.[80]

The NRA's response to the horrors of what had unfolded in the quiet town of Sandy Hook was more guns, not fewer. They backed a 'school shield' proposal – improving school safety by calling for armed guards in every school.[81] 'The only thing that stops a bad guy with a gun is a good guy with a gun,' LaPierre told reporters.[82] One month after the Newtown massacre, the 'NRA: Practice Range' app, which tests shooters' accuracy, came out.[83] The app was originally recommended for four years plus.[84]

The sober reality of all of this is that in the eighteen months before Sandy Hook there were seventeen deaths recorded nationwide in seventeen documented shootings at US schools. In the eighteen months following Sandy Hook, forty-one deaths were recorded in sixty-two incidents: a rise of 141 per cent. There has been a steady rise in the last decade of mass shooting incidents in the US.[85]

So powerless do many Americans feel about their ability to prevent mass shootings that an Oklahoma company has come up with a solution. They sell bullet-resistant blankets to protect schoolchildren – an 8mm pad that they say protects against 90 per cent of all weapons used in school shootings.[86]

What the gun lobby's work amounts to: a nation where schools buy bulletproof blankets for kids.

My attempts to understand the unique relationship with the gun in American culture had led me far across this broad and windswept land. From talking to gun shop owners in Arizona to gun control lobbyists in Washington, survivalists in Arizona and peaceniks in Manhattan, I had sought to understand a little about why the US is so enamoured of the gun. The roots to this lay deep in the soil here, deeper than I could ever dig.

But the trail had led me to the Orange County Fairgrounds in Middletown in New York State. So it was, on a cutting November morning, I pulled my coat hard around my shoulders and walked the length of a massive warehouse that ran parallel to a long line of pick-up trucks and rusting 4x4s. Fifteen dollars and you were in, free to stroll between the yellow signs that rose from the trestle tables. 'Guns wanted,' they said, 'Buying guns, Parts, Ammo.'

It was a county gun fair – just one of thousands held around the US every year, and a family atmosphere hung around it. Fathers ate hotdogs with their sons, grandfathers talked to granddaughters about shooting bears. In one corner a woman with a bullet earring and a camouflage T-shirt was eating, without looking, from a packet of Doritos. A poster behind her read: 'I'd rather be judged by 12 men, than carried by 6'.

Yet, despite the home-loving feel of this place, there was something else that was troubling here. It wasn't the incongruity of the Chinese sellers, with their neatly laid-out tables of laser-sights. It wasn't the M16-shaped BBQ lighters, or the knife called 'The Redneck Toothpick'. These things did not disturb. Rather, it was the table with the spoon on it.

More specifically, it was the table with a spoon with AH engraved on it. Adolf Hitler's spoon, found in 1945 by a US Airborne lieutenant called D. C. Watts, and now yours for $400.[87] The spoon was nestled beside a row of other Nazi objects: a greeting card from Hitler – 'Der Führer des Großdeutschen Reichs'; Eva Braun's calling card.

I looked up and saw another image. This one slightly different; it had hollow eyes and a pockmarked face with drooling lips. A $4.99 Nazi zombie target. I write this because the image stayed with me. Then, as now, it struck me that the Nazi zombie was somehow significant in my journey into the world of the gun. That it was, in a way, the perfect lobbyist in this world – indestructible evil

personified. It was, at least, the perfect reason to own a gun –
combating the zombie apocalypse.

I had seen them throughout gun shows in America: zombie targets,
T-shirts and costumes. There was the Zombie Max bullet.[88] There
were zombie survival camps.[89] *Outdoor Life* magazine even ran a
'Zombie Guns' feature – 'the only way to take 'em out is with a
head shot'.[90]

Dwelling on this zombie metaphor might seem extreme, but in
the US zombie culture is massively popular. The second-season
premiere of AMC's *The Walking Dead*, a series about surviving in the
face of an apocalyptic world, was by far one of the most popular
programmes in the US; Season 4's premiere night later attracted over
16 million viewers.[91] The immensely popular *Call of Duty* has a zombie
mode. Nazi zombies have starring roles in films such as *Zombie Lake*,
Dead Snow and *Zombies of War*.[92] And zombie walks, where people
dress in the clothes and make-up of the undead, have been seen in
twenty countries, with up to 4,000 participants at a time.[93]

This zeitgeist has even had political impact. When in Florida a
state senator in 2014 proposed a bill allowing people to arm themselves in a state of emergency, it was rejected by another senator,
who dismissed the gun-toting bill as 'An Act Relating to the Zombie
Apocalypse'.[94]

The zombie metaphor has been taken up by the pro-gun lobby,
too. In October 2013, hundreds of armed pro-gun rights men rallied
at the Alamo in San Antonio. There, Alex Jones, a controversial radio
talk show host, took to the podium, a semi-automatic rifle across
his back. He outlined a worldwide conspiracy to take away everyone's
guns, calling those who advocated basic background checks 'pathetic
zombies . . . stupid victims that want us to live like they do, slaves'.

Why this obsession with the undead, I wondered? I was concerned
that my thoughts on this would incur ridicule, that the gun nuts
would tear me apart, so I sought higher help.

It came in the form of an academic called Christopher Coker. A
professor at the London School of Economics, Christopher had
written articles about zombies and combat and was happy to meet.[95]
His room was a Wunderkammer of delights, filled with Chinese

Maoist propaganda puppets, Russian dolls with the heads of the Taliban painted on them and West African voodoo dolls. He had taught there for thirty-two years and looked impossibly young for a man of fifty-six. Perhaps luckily for me, he also agreed with my zombie observation. There was something to it, he said.

'West Point cadets read zombie books now; it's penetrated the American military to a remarkable degree.' Outside his door hung a news cutting. The US Defense Department had a disaster preparation document called CONOP 8888. It was a zombie military response document – developed to train commanders preparing for a global catastrophe. The briefing stated, clearly, 'this plan was not actually designed as a joke'.[96]

I asked Coker what it was about zombies that proved so attractive in the world of US guns.

'First of all, you're not dealing with creatures with any moral personalities,' he said. 'You can shoot as many as you like. It's just open fire, and there's no moral conundrum. This has to be seen through the perspective of the War on Terror. Wouldn't it be wonderful if the Taliban and Al Qaeda were actually zombies?'

But he put its attraction as much deeper than just psychotic soldiers and gun nuts. 'Paranoia is extremely important in the gun culture and the NRA . . . America's just paranoid about enemies within. It's always the enemy within. It starts with the Brits, with Benedict Arnold. Whom can you trust? Compare that to zombies – your neighbours can turn into zombies through contagion. That's why you need a gun – to defend your family against your neighbours, essentially.'

Its roots were deep. 'That paranoia comes from Calvinism. And it doesn't matter if you're a Catholic or an atheist, Calvinism deeply permeates the American imagination. There are demons here on earth – they're not waiting in hell. Americans live continually with the idea of fear, and that I think is a very primal fear. It's very much a part of the American psyche. It could be zombies, it could be another AIDS virus. It could be terrorists, of course.'

It reminded me of the 1991 film *Cape Fear*, by Martin Scorsese, when the private investigator Claude Kersek says: 'The South was born in fear. Fear of the Indian, fear of the slave, fear of the damn Union. The South has a fine tradition of savoring fear.'

But it is an odd fear. After all, the general crime rate in the US is on a downward path – 40 per cent lower than in 1980.[97] Knowing this, I could only conclude that these concerns, summed up by zombified danger, were sustained by the ever-present threats that permeate America's news and advertising.

You could see this fear in so much of the gun companies' marketing, too. Glock put out a commercial that focused on the role a gun plays in stopping a stranger knocking on your door and raping you.[98] The hysteria of the new millennium was used to sell guns, with marketing men highlighting the fear that Y2K would cause mayhem: 'dogs through starvation will revert to wild beasts'.[99] And, of course, there existed the perennial concern that Obama was coming to take away your guns, as summed up by LaPierre in an article on the NRA website: 'Obama's Secret Plan to Destroy the Second Amendment by 2016'.[100]

How to protect yourself when the zombie apocalypse comes? Huge caches of guns and ammunition, that's how.

So lucrative is this cultivation of anxiety that the US firearm industry cites it is a key reason to why so many Americans bought guns in 2013. 'Fear of a potential rise in crime contributed to unprecedented industry growth,' one report concluded.[101] Profits flow from terror. The number of criminal background checks for firearm purchases jumped 22 per cent in the month following the Twin Towers attacks. In 2012, following the massacres of Aurora, Colorado, and Newtown, FBI background checks rose some 82 per cent.[102] It was a deep-rooted terror that compelled scared home-owners to keep a pistol under their pillow at night.

The horrible irony is that such an armed response does not actually make you safer. As the *New England Journal of Medicine* concluded: 'Americans have purchased millions of guns, predominantly handguns, believing that having a gun at home makes them safer. In fact, handgun purchasers substantially increase their risk of a violent death. This increase begins the moment the gun is acquired – suicide is the leading cause of death among handgun owners in the first year after purchase – and lasts for years . . . Gun ownership and gun violence [rates] rise and fall together . . . Permissive policies regarding carrying guns have not reduced crime rates, and permissive states generally have higher rates of gun-related deaths than others do.'[103]

Despite this, it remains in the interest of gun companies and the NRA to market the lie that you need a gun to save your life. Manufacturers are faced with the basic truth that, unlike fridges or vacuum cleaners, they sell something that is pretty indestructible. They cannot urge their customers to replace old weapons – the ruggedness of their guns is one of their key selling points. So their advertising has to focus on two other things: the accoutrements that you can buy, like sights and bespoke grips, or the protection a gun offers you from the things you fear.

Adam Smith once said words to the effect that no nation can be happy if the greater part of its citizenry lives in poverty. Perhaps you can add to that: no nation can be happy if they live in poverty or fear, or both. What I had seen on my travels was that the one thing that truly, consistently transformed poverty and fear into deadly violence was the availability of guns. And this easy availability of guns was largely down to two groups: the lobbyists who secured the political will for their sale and, of course, the people who made them. The manufacturers.

15. THE MANUFACTURERS

Guns as the first product in history – revelations on manufactured death – the workings of a pistol factory on the Black Sea coast in Turkey – massive gun profits and political influence in New York, USA – gaining access to the hell-dog boardrooms of the world's largest gun manufacturer foiled

November 1851 was, by any stretch of the imagination, a remarkable month. It saw the publication of *Moby-Dick* by Herman Melville; a book seen by some, with its monumental ambition, as the first modern novel.[1] It was the month that celebrated the laying-down of the first protected submarine telegraph cable, on the bed of the English Channel, foretelling an era of global communications.[2] And, on a London day when the sky at dawn had been the colour of the Thames but was now brilliant and blue, it saw Samuel Colt standing before the bewhiskered and besuited members of the Institution of Civil Engineers in London, waiting for them to fall quiet. He had something to show them.[3]

Like many American engineers, Colt had travelled to an England entranced by the displays of the Great Exhibition, and it was there that he planned to show off his design.[4] What he presented fascinated the crowd: an invention with interchangeable parts.[5] It was the Navy Colt Revolver. What captured the greater imaginations of these straight-backed and high-browed engineers, though, was that 80 per cent of the gun had been made on machines: a revolutionary departure from crafted metal gun parts traditionally lathed by hand.[6]

By the time that Colt had finished talking, many in the room were won over to this way of mass production.[7] The awards board presented Colt with the prestigious Telford Gold Medal, and already the use of this manufacturing process to make guns was spreading from the Connecticut River Valley, where the majority of American firearms were once made, across the US and beyond.[8] This production method, defined by its extensive use of inter-changeable parts and mechanisation to produce them, became known as the 'American System'. It was soon to be the central way to mass-produce so many things that define our modern life – cars and bicycles, clocks and furniture. And, of course, guns in their millions.

It is no coincidence that Remington once made typewriters, sewing machines and cash registers.[9] Or that Winchester made handcuffs, dishwashers and toilet flush valves.[10] No surprise that the inventor of the Kalashnikov rifle once dreamed of designing agricultural machinery,[11] or that the gun maker Glock started off his life designing plastic curtain rod rings.[12] The principles of production were all the same – it is just that guns have a very different use than other mass-produced items.

Colt was at the forefront of this revolution in production. For that reason some have even called his Navy Colt Revolver 'the very first product in history'.[13] And mass-produced it certainly was. Over a quarter of a million revolvers were made, and they soon could be found from the plains of the US to the Russia steppes, from the far tropical reaches of the British Empire to the dusty plains of the Ottoman Empire. It was a design that launched the modern gun industry – enabling firearms to be made in their tens of millions. Colt alone has sold 30 million guns since its founder filed his first revolver patent.[14]

The number of guns produced today is disturbingly huge. Over 1,000 companies from some 100 countries produce guns and ammu-nition.[15] The exact figures are hard to come by, but the oft-repeated estimate is that about 8 million guns are made every year.[16] This might be erring on the low side. Gun makers from just the US churn out nearly 6 million guns annually, and there are countless weapons made in secretive factories in China and Russia whose production we never catch a glimpse of.[17]

What we do know is that as many as 100 million AK-pattern weapons have been produced since the 1950s, and about 12 million AR-15 type rifles have been made since the 1960s.[18] We know that as many as 17 million Lee Enfield–series rifles and some 7 million G3-pattern rifles have also come off factory lines across the world.[19]

The US accounts for at least half of the world's small-arms production, with seven of the world's top ten gun producers based there.[20] Three American brands – Ruger, Remington and Smith & Wesson – each made over 10 million guns between 1986 and 2010, about 40 per cent of US domestic gun production.[21]

Europe is the other major producer. Italy, Britain and France have some of the world's largest gun manufacturers. On average, Beretta makes 1,500 weapons a day.[22] And, of course, there are Russia and China, as well as emerging gun producers such as Turkey.[23]

What is clear is just how much money can be made from all of this. In the US, in 2012, the industry tallied up almost a billion dollars in profits.[24] In 2013, Ruger had sales of almost $700 million.[25] The figures are so large that the US gun industry boasted they were a 'bright spot in the economy', having witnessed 'two decades of steady increases in gun sales, including five years of record growth'.[26]

Unlike ballistic missiles and attack aircraft, the technical barriers to making a gun are incredibly low. The granting of licences and the spread of technology mean that companies can produce guns without expensive research-and-development programmes, estimated to be just 1 per cent of the turnover of more established gun companies.[27]

This low level of investment has meant that some gun makers have become inconceivably wealthy. Gaston Glock is a man so rich he reportedly could buy a $15 million horse for the Glock Horse Performance Centre, run by his wife, Kathrin, a woman about five decades his junior.[28] The gilded offspring of the Beretta family get sycophantic vanity-pieces in magazines such as *The Billionaire*.[29] Even the infamous banker J. P. Morgan made some of his first profits selling faulty guns.[30]

Of course, it is not all plain sailing. The company that made

Kalashnikovs endured hard times after a collapse in orders followed the fall of the USSR.[31] A flood of knock-offs and poor management drove it into the red, with losses of $50 million.[32] Today the iconic gun manufacturer Colt lurches from financial crisis to financial crisis.[33] So gun manufacturers, with their lobbying and their marketing, need to continually encourage people to buy more and more of their product. It's a remarkably volatile market that demands, as I had seen, the marketing of fear and the production of new types of weapons to keep it afloat.

Innovation is one such thing that some makers turn to in order to generate a profit. Helmet systems have been developed so that soldiers can now point mounted guns by just moving their head. There has been the creation of the 'supergun' – where target-locking technology can turn any rifle into an ultra-accurate sniper's weapon. There is the geo-location system developed by Yardarm Technologies that can remotely fire or lock a gun. And there is the development of weapons that are unnecessarily extreme. One gun system called the Firestorm has a firing rate of 20,000 rounds a minute.[34]

There is also the development of ammunition. From non-toxic ammo[35] to the creation of guided smart bullets,[36] a wide array of vicious forms of munitions are produced. There are shotgun shells that convert your gun into a short-range flamethrower;[37] or 'Subsonic Controlled Fracturing bullets' that 'quietly exit the barrel of your firearm, enter your target and then via hydraulic pressure, fracture into razor sharp petals and a base'.[38]

Looking at these inventions makes you wonder what sort of people lie behind them – what motivates them and how they justify their work. And so, nearing the end of my journey, I turned my attention to the gun makers.

Two emails pinged into my inbox in quick succession.

The first was from A. V. Rockwell, a filmmaker from New York.[39] She had been the director of the music video that suggested Paris

Lane had killed himself because of the threat of retribution from some gangbangers he had ripped off. Her email read differently from her film.

> Although the video was inspired by Paris' story, the majority of the events were fictionalized. According to his close friends, no one knows for sure why he committed suicide, but he had no immediate reasons to fear for his life. He did however, have a troubled upbringing. I hope this helps, sorry to have stirred you in the wrong direction.

So that was it. Not a gang-related thing after all. Just an act of lonely desperation made deadly with a loaded gun. I had been distracted by a manufactured reality – one where suicide was turned into a form of manslaughter and where a young man's 'pride' was kept alive. Perhaps suicide is too ugly a thing to sell records off the back of. In the end, it was another pointless death down the barrel of a gun.

Then the next email arrived. I had written to the top twenty gun manufacturers worldwide asking if I could have a tour of one of their factories. Smith & Wesson wrote back. 'The information you seek is largely proprietary, so I am afraid we are unable to help.' Blaser, the German hunting rifle company, were, at least, honest when they said: 'We have to be wary of having our name published in a context which could lead to negative association.' The others just didn't return my email.

But this one was different. It was from a Turkish gun maker called Utku Aral. 'Thanks for your kind mail and interest to our company,' he had written. 'You can be our guest . . . I will be waiting information from you.'

'Any spare time in Samsun may well weigh heavy on a traveller's hands,' was the guidebook's damning assessment of the Turkish city of over half a million that lies on the shore of the Black Sea. They had a point. It was a functional city whose tourist office had long since shut down and where, even in summer, light rain and dark puddles marked the grid of its leaden streets.

Nurey, my driver, did not speak English, but as he drove me to the gun factory he pointed out the few sights there were to see. This port city had once been a trading station for the Pontic kings, the Romans and the Genoese, but there was little to show for it. There was a dull glistening statue of Atatürk – the father of modern Turkey – under an ashen sky. The great man's boat, the SS *Bandırma*, lay further along, flanked by a row of listless red crescent flags. It gently swayed in a faded harbour speckled with the distant black rectangles of silent container ships.

We drove on, through roads slick with a thin spread of mud, passing women in tie-dye hijabs darting to avoid the rain. Above us rose crumbling concrete minarets and the silhouetted ticks of off-black birds caught against a clouded sky.

Then, after a seeming endless row of small shops selling tyres and windscreen wipers and chrome hubcaps, we passed a sign: 'Samsun Organize Sanayi Bölgesi'. This was the industrial park where Utku Aral's factory lay. He was the CEO of Canik, Turkey's leading pistol manufacturer, with a production line of over 80,000 sidearms a year.

We pulled into the grey courtyard, and I was shown up to Utku's office. He was not there, so I did what journalists do – and had a nose around his office. The walls were lined with thin shelves, each one filled with mementos from his business trips: a gold man statue from Kazakhstan, a Spanish bull with a bandana collar, trinkets from Israel, the US, Pakistan – a reflection of the twenty countries where Canik's pistols were sold. Besides them were framed photographs of Utku – dark-suited and serious, displaying his pistols at a gun show, flanked by men in high military caps, and fierce and thickset politicians. I wandered around. On his desk was a picture of a young wife looking serene in a ball gown.

Utku came in. Wearing a white button-down shirt and chinos, at thirty-two he was already managing director of one of his family firms. With a degree in mechanical engineering from a top university and a corporate lawyer wife, he possessed the self-confidence of a man who was assured of his place in the world. He settled down into a deep leather chair and called for small cups of sweet Turkish tea. They arrived, softly clinking, and, in perfect English, he began to explain a little about his industry.

Canik was a mid-sized Turkish gun company – the thirteenth-biggest defence manufacturer in Turkey – with a turnover of about $40 million. They produced semi-automatics and sniper rifles, but the main weapons they manufactured were handguns.

He chose his words carefully, a trait echoed in his neat haircut and precisely ironed shirt. But he needed to be cautious. His company had been born from a government initiative – the East Black Sea Weapon Project. Even now the government had tight controls on him, with frequent checks and endless permissions required for exports and production.

Having spent time in the Turkish army, he spoke of patriotism. He felt making pistols for the Turkish police force was part of his duty to protect his nation. 'The volatility of Afghanistan, Iraq, Syria, and all of our borders', he said, in a singsong list, it bothered him. 'Do we know who the Syrian refugees coming over our border are? What if they are IS?' he said, referring to the Islamic militants fighting a twenty-hour drive from where we sat.

This belief in the right of self-protection led him to sell weapons to Cameroon, Burkino Faso, Thailand, Zambia, South Africa, Jordan, Kazakhstan, Tanzania, Peru and Chile.

'If a CEO sits at home, they are not doing their job,' he said. Last year over 100,000 Turkish pistols were sold into the US. The long arm of the Second Amendment was there again: 'If the US says "no more" to gun imports, then small-arms manufacturers all over the world are nothing.'

I asked about corruption in the arms trade. He was surprisingly frank. 'Corruption exists in our sector as it exists everywhere. Whoever says that is not the case is lying.' The problem, he said, was when governments were involved in the deals. 'Yesterday I had a meeting over Libya. There is a UN arms embargo on Libya at the moment, so we can't export there. But Russia, China – they don't care about UN decisions.'

I wondered if this was just smugglers taking advantage, and he shook his head. 'Everything is done by governments. Taking guns from someplace to somewhere else without control can't easily be done. It has to be done by a government.'

He explained how things work. 'We go to Ghana and we say we want to sell them guns. But Ghana asks for a soft loan from me and says if I give this loan to them they will buy my guns.' This is how it is – he said – governments secure these big arms deals but then finance them through the back door. 'This is the game.'

We spoke for hours, and then, sensing I was keen to see the guns in action, he led me downstairs and out to the testing range. At that moment his biggest buyer was the Turkish police, ordering some 30,000 pistols from him a year, but that contract was hard to win. The police take his pistols and test them, drop them onto concrete from a height, put them into deep freeze and ovens, and then do what is called a 'torture test' – leaving them in a salt bath for twenty-four hours and then firing 10,000 rounds from them in a row. They fire 30,000 more rounds if the pistols are being sold to Turkey's Special Forces units. The men who do the testing are so muscle bound from shooting they can cock their pistols with the strength of their arm movements alone.

He handed me one of Canik's pistols and showed me how to stand – legs apart, fixed arms, hips to the rear. I lifted the gun and fired ten steady rounds at the quivering target. The firearm was heavier than I had imagined it would be. By the end the sights were shaking in my hand.

My aim was poor. Years before, I had been able to get tight group-ings when down the range, but now I was off. Perhaps I was just out of practice. Or perhaps I had lost interest in giving the gun my best – I had seen so much of the harm it had wrought that shooting it became a chore, and each sharp crack of the pistol brought back memories.

Utku took me back to the main building. The low hum of machinery had been a constant throughout our conversation, but, as he ushered me down a corridor to a door leading into the factory floor, the noise grew. Throbbing and vibrating machines stood on all sides – huge square mechanical beasts from Germany in black and grey, tended by sullen men in blue polo shirts. Utku employs 245 people – around two-thirds here on the production line.

'Every second is money,' he shouted above the noise. The forty-eight machines operate six days a week, and for five of those days they are on all the time. 'Now you can do 98 per cent of the same job on one machine. It might cost $384,000, but it does the same job as twenty-five different machines,' he said. Peering inside, you could see metal being lathed into barrels and stocks. Beside them lay thousands of finished parts. This truly was mass production in action. 'Small-arms production is the automated industry of the defence market,' shouted Utku.

We walked down to a storeroom – there thousands of specialist tools for cutting lay in neat, labelled rows. Tools from Germany, Israel and Italy, some costing as much as $1,100. It struck me how the gun industry doesn't just affect groups like lobbyists and health-care units and morticians. It has a hidden financial impact on the manufacturing world.

'If there is no defence industry there is no tool industry. If there is no tool industry there is no machine industry,' Utku said. It was big money, too. He opened up one cabinet. 'That whole cabinet,' he said, leading me onwards, 'was worth $110,000.'

We came to other tools needed to produce the gun. Here was the gauge room; on all sides stood fine-tuning devices from England. Utku explained, 'Without gauges you cannot have the same perform-ance for every gun. If you don't have gauges then they cannot ask for a spare part – you need to make 2,000 exactly the same.' The lathes here cut to an exactness of 0.01mm. 'Precision,' he said, 'is everything.'

I asked him how guns had changed him in this regard. 'It's made me a control freak,' was his answer. And it struck me that here everything was about control and detail – but what was produced could let loose such chaos and anarchy. I guess Pandora's box had neat lines and straight corners, too.

I asked Utku about this – about the fact that the things he makes must kill. 'I don't care about this,' he said with a forcefulness that surprised me. 'If human beings were not the most dangerous thing in the world it would be different. If we could have a constant civilisation in the world, then gun control might work. But everything is not getting better in the world.'

His guns may well have played a part in that. He had once been contacted by Interpol to explain why one of his pistols had been used in a murder in El Salvador. It turned out it had been smuggled there from Guatemala. But here in this broiling and bustling factory it was hard to imagine the path these guns would take, the lives they would decimate. Here guns were just a product – benign, unfinished. They posed no threat.

Yellow fork-lift trucks busied around the painted demarcations on the floor, passing stacks of unfinished slides and trigger guards waiting to be dipped in chemicals or hardened in huge metal furnaces. Utku led me upstairs to one final room – where the guns were assembled. I walked in, and memories of Brazil came quickly back, because here the walls were lined with wooden slats. In each pigeonhole lay a new pistol, as yet untainted by the mark of Cain: hundreds upon hundreds of them. On a table lay handfuls of other handguns – in graphite black and chrome, in general police-issue slate and in 'fancy Arab' gold plate. They were called the Shark, the Piranha, the Stingray.

Here, in this wall-lined room of guns, I felt I was coming close to the end of my journey. I had been to places where I had seen the need for guns to keep the peace. I had been to places where the gun only seemed to disrupt things. I had met whole communities who gathered without bloody incident around the gun and its use, and yet had also seen cemeteries filled with the graves of those killed by them. For each truth I alighted upon, another seemed to run counter to it.

Utku spoke. 'I believe it is not possible to control guns. We are animals. When we are poor we are worse animals. Humans don't care about others. They want their own success, that's all. To stop gun violence is just a dream. Guns are a very necessary evil.'

Perhaps he was right, but I did not want to believe in such a bleak view of the world. I wanted to feel that good comes from making a stand; doing nothing was part of the problem. We should, if anything, confront evil face to face.

I was back in New York. I had one final thing to do: to meet the men whose investments backed these gun manufacturing companies. To chase the money to its end.

The building at 875 Third Avenue rises anonymously. It has the discreet, banal architecture of a thousand other bland officer towers just like it. Beneath, in its bowels, there dwell the sort of shops you would expect in any large New York office building. A sushi bar for those alpha males who want to stay slim. A Baskin Robbins ice cream parlour for those not so alpha. A Subway for the rest. But the shop that caught my eye was the one that sold edible arrangements – fruits dressed up as flowers. I wondered whether some of these flower-fruit-themed baskets had been delivered to the board-rooms of the tenth, eleventh, twelfth and fourteenth floors that rested in chrome and glass, high above me. Whether there stern men in suits, discussing profits and losses and balance sheets, may have paused for a moment and, with poised fingers, plucked at a ripe strawberry from a 'Delicious Daisy®'. Then, mouths savouring the lightness of the fruit, I wondered whether they went back to talking about how the latest gun massacre could have a positive impact on sales. Because high above me were the offices of Cerberus Capital Management.

Cerberus are a big-player private investment firm. 'Dedicated to distressed investing', they say. Dedicated indeed – they have over US $25 billion of investment on their books. They deal in things like 'non-control private equity', 'distressed assets' and 'corporate mid-market lending' – all of which are carefully constructed words that hide the hard realities of what this company actually does.

What is more tangible than these oily words is that, in April 2006, Cerberus Capital Management had decided to get into the gun business in a big way. Their first major purchase was for the semi-automatic rifle maker Bushmaster Firearms. Cerberus took them on for around $76 million.[40] The next year the asset group formed what they called the Freedom Group[41] and set upon a gun company buying-spree, snapping up Remington and a slew of other firms including ammunition, silencer and body armour makers.[42]

Today, by Freedom Group's own count, they are the world's largest manufacturer of commercial firearms and ammunition.[43] In 2012 they made over $1.25 billion, of which about 60 per cent was just from selling firearms. The next year they sold 1.8 million firearms and 3.1 billion rounds of ammunition.[44]

I walked back upstairs and checked my phone for the fifth time that day. Nothing. The PR company that handles Cerberus Capital Management's media – Weber Shandwick – had not returned my call.[45] I was not surprised, mainly because Cerberus's CEO, Stephen Feinberg, once reportedly said at a shareholder meeting: 'If anyone at Cerberus has his picture in the paper and a picture of his apartment, we will do more than fire that person. We will kill him. The jail sentence will be worth it.'[46]

The entire gun industry is shrouded in secrecy. Only a small fraction of gun companies are publicly listed on stock exchanges, and most are not obliged to publish detailed accounts or annual reports. All but one of the US major domestic gun manufacturers – Ruger – are privately held companies. And there hasn't been a whistleblower in the gun sector as seen in other industries such as the tobacco, pharmaceutical or financial sectors. Some have even called the gun industry 'the last unregulated consumer product'.[47]

The handy thing about Cerberus and the Freedom Group, though, is that they are one of a few who have actually published their accounts.[48] These reveal that the Freedom Group sold 400,000 more guns in 2013 than they did in 2012 and that their 'work in shaping International requirements' led to 'an estimated $50 million carbine contract with the Republic of the Philippines' – a country where about a fifth of the population live beneath the poverty line.[49]

It was enlightening reading, but I had a very specific set of questions I wanted to ask Feinberg, so I walked over to the reception desk computer and typed in his name.

'Sorry, there is no further information listed under Stephen Feinberg,' flashed up the answer.

I did this for a number of Cerberus employees until the security guard, fierce and suspicious, glared at me. 'Is there anything I can help you with, sir?'

I explained I was a writer and had been trying to get hold of Feinberg, and he looked even more furious. I walked away, his eyes on my back. I called the PR company – no reply. Perhaps they rarely reply to journalists calling about guns. After all, Weber Shandwick is one slick PR company, which lists clients such as the Colombian government,[50] the US army and BAE systems,[51] the sort of company that hires people who write articles called 'Reputation Warfare'.[52] It's certainly reputation warfare that Cerberus is forever waging.

The asset company has had its fair share of PR disasters. There was the matter of the mass recall of a Bushmaster Adaptive Combat Rifle because the semi-automatic function could turn into a fully automatic one.[53] Then there was the time when CNBC aired the documentary *Remington Under Fire*. The reporters looked into allegations that the Remington Model 700 rifle had an unsafe trigger that could cause accidental discharges, reportedly leading to 'multiple deaths and hundreds of serious injuries'.[54] The company called the claims 'baseless and uproven'.

But their biggest PR disaster was Sandy Hook. The weapon used by Adam Lanza in the Sandy Hook Elementary School massacre was a Bushmaster AR 15 rifle – the one they marketed with the line: 'Any gun will make an intruder think. A Bushmaster will make them think twice'.[55]

It was also not just Lanza. The DC snipers who haunted Washington's beltway did so with a Bushmaster, using it for eleven of the fourteen shootings,[56] while William Spengler, the killer who shot four and killed two volunteer firefighters responding to a fire in Webster, New York, in 2012, was also said to have used one.[57]

But there was another concerning thing: Cerberus owns a health-care company. It is called Steward Health and has 17,000 employees serving over 1 million patients in New England,[58] which means that, with almost 46,000 cases of violent crime in New England in 2011,

Cerberus almost certainly runs a business that has to treat gunshot victims.[59]

Think on it. The biggest firearms company in the world, one that makes immense profits selling guns, some of which are used in mass shootings, is also a company that seeks to profit from treating the victims of gunshot wounds.[60]

These ugly truths caused such an uproar that, in December 2012, in response to the Sandy Hook massacre, the company announced it would begin selling its investment in the Freedom Group.[61] Perhaps it was Stephen Feinberg's father, who actually lives in Newtown, who urged his son to get rid of the companies.[62] Perhaps it was the pressure from investors such as the California State Teachers' Retirement System (which had invested $600 million in the Cerberus funds) that sparked the announcement.[63]

Or perhaps it was just PR talk. Cerberus never did what they said they were going to do. A year after Sandy Hook the company that made the guns that killed those kids declared its profits. They had made about $240 million – a 35 per cent rise in Freedom Group's earnings on the year before those children died.[64]

I had looked for the words 'Sandy Hook' in the Freedom Group's annual report. Nothing. Instead it had phrases like 'amortize actuarial gains and losses' and 'supplemental financial metric for evaluation of our operating performance'. No Sandy Hook. But there was mention of $3.4 million of state and federal tax credits. US taxpayers, it seems, subsidise the manufacturers of US guns.[65]

Above me then, inaccessible, were their clean and well-lit offices. In there sat Cerberus's CEO, Stephen A. Feinberg.[66] A fifty-four-year-old father of three, he reportedly keeps an elk's head on his wall and rides a Harley. He calls himself 'blue collar' yet brings home as much as $50 million a year, has an apartment on the Upper East Side and went to Princeton.[67]

But I was not surprised that Feinberg would think of himself as just an ordinary working guy. His company's words are similarly oblique. 'We are investors, not statesmen or policymakers,' their statements read.[68] Yet money in the US is always a gateway to politics and policy, and Feinberg certainly has a key to the hearts of the

Republicans. He has donated over a third of a million US dollars to them in the last ten years. And, for a man who does not see himself as a policymaker, he's pretty selective which politicians he gives his wealth to: men like Utah senator Orrin Hatch,[69] a man with an A+ rating by the NRA for 'opposing any international treaty by the United Nations . . . that would impose restrictions on American gun owners',[70] or Montana's Max Baucus, one of four Democrats who voted against the amendment to extend background checks to private gun sales – and whose vote helped kill the bill.[71] So of course he's an investor, not a policymaker. Just as his wife donating to the National Republican Congressional and Senatorial Committees and a host of other NRA A+ rated politicians is also in line with Cerberus's humble position that he's just an apolitical investor.

Perhaps Feinberg's most direct link to the pro-gun Republican caucus, though, is the one-time vice president of the United States. J. Danforth Quayle is chairman of Cerberus Global Investments.[72]

Then there is George Kollitides – chairman and CEO of the Freedom Group.[73] Unlike Feinberg, whose online presence amounts to a fleeting picture of a mousy-haired man with a small moustache, Kollitides's online profile reveals a life of self-regarding pleasure. At forty-three, he has the trim appearance of a military man, hair shorn hard at the sides, a face showing more pride than humour. Photos catch him posing with the antlers of downed stags[74] or crouching behind a freshly killed bear.[75]

His wife, Karen Kollitides, blonde, groomed in a specific way, is captured in the glare of high-society events such as 'Models 4 Water', a charity that provides clean water to remote parts of Africa.[76] It makes you wonder whether the guns her husband produces have been involved in displacing refugees in Africa and causing untold families to eke out dry-mouthed lives in endless scorched deserts.[77] Certainly his company's guns have been found in the hands of militants murdering for the Islamic State.[78]

Like Feinberg, George Kollitides is also a major donor to politics and lobby groups. In December 2013, the NRA inducted him, along with the Freedom's Group vice chairman and president, into the 'Golden Ring of Freedom', a group of individuals who have given the NRA at least $1 million.[79] He has also recently made political donations to plenty of the Association's top-rated Republicans.[80]

If you want me to believe these men give their money freely, without agenda, to the US political elites – that they are just humble investors – then I guess I might as well believe that guns have nothing to do with killing people as well. Just as I might as well believe that Wayne LaPierre's comments about media manipulation and undue political influence does not also happen at the NRA.

Not surprisingly, then, neither Feinberg nor Kollitides nor LaPierre would speak to me. So I left those pristine, well-lit, comfortable offices and wandered outside.

Cerberus. It was called after the three-headed dog that guards the gates of Hades. Feinberg apparently liked the idea that one of the dog's heads was always on watch, just as his firm would guard its clients' investments around the clock.[81] For me, the name had different overtones. After all, part of me felt that I had been to parts of Hell. I'd heard the cruel silence of Sandy Hook. I had seen the horrors that American guns had wrought in Honduras and Mexico. I may have even been shot at by one of Cerberus's weapons.

Now, an age since setting off on this journey, I had come up against Cerberus – the dog that guarded the entry into Hell – and it wasn't letting me any further.

16. THE FREE

Freedom and a world without guns considered

The Statue of Liberty rose above me, a chlorine-green giant, a silent presence. People walked around its base, glancing up at it, but it dominated that small island so much that people focused on little things. Mainly on each other, for here was the world: teenagers flirted, children ran and called into the drifting wind. From east to west, north to south, it was, as it had been for so many years, a meeting place of nations.

I had come here after the stonewalling of Cerberus. If the dog at the gates of Hell wasn't going to talk to me about the inner workings of the Freedom Group, then a ferry ride to the most famous symbol of freedom in the world seemed an appropriate response.

So there I stood, gazing at the swooping gulls, and felt the Atlantic wind pick up. The New York skyline twinkled on the horizon. I looked up at the statue. They had called her 'Liberty Enlightening the World'; she was the 'Copper and Iron Colossus' whose carapace was as thin as two American pennies. Their choice of metal – copper, not bronze – was deliberate. The French designer, Frédéric Auguste Bartholdi, did not want a statue 'cast from cannon captured from the enemy'. The European military tradition had been to create their victory statues from the bronze guns of the defeated, but copper was the metal of coin and commerce. She was almost a symbol of liberty from the repression of the gun by Europe's military and political elites.

Across the dancing waters lay the endless spread of America, and I thought how much things had changed. Today Lady Liberty still stood for freedom – but it was a different form of liberty than that which Bartholdi had perceived. Now it was wrapped up in the logic of the American Second Amendment's right to bear arms: a right conceived in an age of simple single-shot guns, not ones that could decimate a school playground in the blink of an eye. Any questioning of this right came up against a barrage of opposition – a stone wall of lobbyists, manufacturers' political donations and silent moneyed men.

It was a logic of the right to bear arms that had spread far from America. US guns ended up fuelling drug wars in Central America and Mexico, and vicious conflicts in the Middle East and beyond. Second-Amendment-quoting lobbyists had hobbled international attempts to address the spread of illicit guns worldwide. And the logic of American mass production meant that far more guns were being made than were ever being destroyed.

I had come to the end of my journey. Behind me lay a battered landscape of memory – the worlds of pain, power, pleasure and profit. What I had seen was that the gun's impact on lives – our lives – was divided into dozens of different realities. That communities living with guns at their epicentres often lay far removed from other communities with other guns. Gun lobbyists never got shot at, American hunters don't meet Salvadoran gangland killers. Gun makers focused on the minutiae of a barrel's width, while doctors frantically focused on stemming the blood from the imprecise holes caused by a bullet's spin.

This divided world was the root of the gun's hold over us. We could never get rid of its ability to impart pain, because doing so meant taking away someone else's power, their pleasure and profit. Those who say guns don't kill people, that people kill people, just haven't seen the whole picture. They have only seen one element of its transformative power. Yet I had seen all of the varied faces of the gun, and to me it was unequivocal – guns kill.

Transformative: this was the essence of what guns were. They took man's basic impulses and stretched them, from the pinnacles of empowered wealth and desire to the depths of pain and war. They could turn an argument into a deadly confrontation. Make you give all your attention to a man you wouldn't, shouldn't give a second thought for. Save you and sink you. Of course the gun gives us freedom, I thought. Freedom to do as we want – or for someone else to do as they want to us.

That was the horror.

From the tomb-like arsenals of Brazilian armouries to the sun-touched mountain heights in South Africa, I had seen guns transform situations, people, ideologies and even me. It had left me on edge, spread thin. I felt fearful – death had left its secret mark; wars don't end just because you are not there. And I feared the future. I anticipated the criticism this book might bring – angry words from those wedded to their right to own a gun.

But I had seen what I had seen, been marked by it, and it felt as if words were the only thing I had left. And yet that fear somehow left me, for a moment, here. I looked up at the looming Statue of Liberty and down at the worn grass that surrounded her. On one patch some central European teenagers were sunbathing. A man sat on another, slowly unwrapping a silver-foiled sandwich. Here was peace.

A thought struck me: despite the inalienable right to own a gun in the US, you cannot go to this emblem of America armed. Liberty Island is a federal property, and National Park rulings ban all weapons. Tourists join long, slow queues on the New York shore and are herded through airport security scanners to make sure no guns are brought here. This has created an island that has virtually no crime. The United States Park Police were unequivocal: 'We located no statistics of any firearms incident on the Statue of Liberty National Monument from 2000 to the present.' This, despite about 20 million visitors travelling there over that time.

It was possibly the safest public space in the world.

I looked around at this artificial vision of freedom and breathed out slowly. What the world might be if we had no guns at all, I thought. And then the sun flared, and the light grew and filled the sky.

What the world might be.

NOTES

Preface

1. The collated data for the figures listed here can be found at http://www
.gunbabygun.com/gun-baby-gun/massacres-like-in-oregon-are-just-the-tip-of
-the-iceberg-of-why-america-has-a-gun-problem/

Chapter 1: The Gun

1. In 2012 in the whole of London, a similar-sized city, there were less than
100 murders. If you look at Brazil's death rate in terms of its ratio, it has the
fourth-highest rate of gun deaths in the world – about 19.3 per 100,000
people. It's estimated about 95 per cent of these are homicides.
2. In a study conducted by the NGO Viva Rio, 11 per cent of the 10,549 guns
seized from criminals in Rio de Janeiro state between 1998 and 2003 once
belonged to the military police.
3. The Small Arms Survey estimated a figure of 875 million guns, but this
was in 2007. As more guns are produced than destroyed every year, almost
a billion guns seems a fair estimation. Other facts from: http://www.small
armssurvey.org/fileadmin/docs/A-Yearbook/2007/en/Small-Arms-Survey-
2007-Chapter-02-summary-EN.pdf; http://www.oxfam.org/en/pressroom/
pressreleases/2012-05-30/ammunition-trade-tops-4-billion-yet-little-
regulation-control-and; http://www.smallarmssurvey.org/weapons-and-
markets/producers.html; http://child-soldiers.org/global_report_reader
.php?id=562/; http://allafrica.com/stories/200706140975.html
4. http://www.culture24.org.uk/history-and-heritage/military-history/art
30537
5. Maxim was, of course, an American resident in Britain. As for the mechanism
of firing: in most instances, it starts inside the bullet, where a propellant, such
as gunpowder, lies. Ignite this and you produce a force. As this propellant

burns, gases are released that generate intense pressure. This pushes the bullet down the barrel of the gun – the thing that gives a gun its 'bang', like the uncorking of a bottle. As the bullet flies through the air, gravity and air resistance reduce the bullet's speed and trajectory. Over short distances, the bullet more or less travels in a straight line. But over bigger distances, the bullet flight path curves downwards. The more streamlined and pointed the bullet is, the more easily it will travel through the air.

6. It was a weapon that had had some of its initial military trials in Sandy Hook in the US – a town later to become synonymous with another rapid-firing rifle used in a horrific elementary school killing.

7. Ian Fleming initially armed his spy with the .25 ACP Beretta Modelo 418.

8. http://www.smallarmssurvey.org/fileadmin/docs/A-Yearbook/2007/en/full/Small-Arms-Survey-2007-Chapter-02-EN.pdf

9. Ibid.

Chapter 2: The Dead

1. http://www.genevadeclaration.org/measurability/global-burden-of-armed-violence/global-burden-of-armed-violence-2011.html

2. http://www.genevadeclaration.org/fileadmin/docs/Indicators/Public_Health_Approach_to_Armed_Violence_Indicators.pdf

3. http://www.unodc.org/documents/gsh/pdfs/2014_GLOBAL_HOMICIDE_BOOK_web.pdf – this report states that firearms account for four out of every ten homicides at the global level but offers no data to back this claim up. The Small Arms Survey estimates that 90 per cent of deaths in conflict are from guns, so saying five out of ten homicides are gun-related seems a reasonable estimate.

4. http://www.jhsph.edu/research/centers-and-institutes/johns-hopkins-center-for-gun-policy-and-research/publications/IPV_Guns.pdf

5. http://www.globalissues.org/article/78/small-arms-they-cause-90-of-civilian-casualties

6. Interview with World Health Organization suicide researchers.

7. M. Peden, *Second Annual Report of the National Injury Surveillance System.* Violence and Injury Surveillance Initiative, South Africa: October 2001. Cited in: http://injuryprevention.bmj.com/content/8/4/262.full#xref-ref-5-1

8. http://www.theguardian.com/news/datablog/2013/sep/17/gun-crime-statistics-by-us-state

9. http://www.theguardian.com/news/datablog/2012/jul/22/gun-homicides-ownership-world-list; http://pt.igarape.org.br – over 34,000 people killed by firearms in one year has been recorded, three times that of the US gun homicide rate.

10. Violence Observatory at the National Autonomous University of Honduras

(NAUH). Across Honduras, between 2005 and 2012 there was a 93 per cent increase in the homicide rate – from 46.6 per 100,000 to 90.4 per 100,000: https://www.unodc.org/unodc/en/data-and-analysis/homicide.html

11. Honduras is 283 per cent worse than the rest of Latin America and the Caribbean. There were 7,172 homicides in 2012 there. This is a rate of 90.4 per 100,000. The average rate that year in Latin America and the Caribbean was 23.6: http://aoav.org.uk/wp-content/uploads/2014/06/Crime-and-Violence-in-Latin-America-and-the-Caribbean.pdf; US deaths stood at 5.2 in 2011: http://www.cdc.gov/nchs/fastats/homicide.htm

12. http://www.theguardian.com/world/2013/may/15/san-pedro-sula-honduras-most-violent; data shown to me by the San Pedro Sula morgue showed there to be 1,971 homicides in the region of San Pedro Sula in 2013 – data published on www.gunbabygun.com

13. http://www.newrepublic.com/article/119026/guns-fueling-immigration-central-america-come-us

14. This is for 'Regional San Pedro Sula' – data from the Ministerio Público and posted on http://www.gunbabygun.com/gun-baby-gun/san-pedro-sula-data-dangerous-city-earth/

Chapter 3: The Wounded

1. http://www.propublica.org/article/why-dont-we-know-how-many-people-are-shot-each-year-in-america – these numbers include only injuries caused by violent assault, not accidents, self-inflicted injuries or shootings by police; http://www.fbi.gov/about-us/cjis/ucr/crime-in-the-u.s/2011/crime-in-the-u.s.-2011/tables/expanded-homicide-data-table-8

2. http://www.plymouthherald.co.uk/Seven-people-treated-gunshot-wounds-Plymouth-s/story-22761113-detail/story.html

3. http://www.gunpolicy.org/firearms/region/united-kingdom

4. http://www.childrensdefense.org; the comparison numbers work out at 32,223 in Iraq and 15,438 in Afghanistan (from US military personnel wounded in action in Iraq and Afghanistan as of 5 March 2012; Centers for Disease Control and Prevention. 2008-2009. 'Fatal Injury Reports.' Accessed using the Web-based Injury Statistics Query and Reporting System (WISQARS). US Department of Health and Human Services).

5. Marie Crandall et al., 'Trauma Deserts: Distance from a Trauma Center, Transport Times, and Mortality from Gunshot Wounds in Chicago', *American Journal of Public Health*, 103, 6 June 2013, pp. 1103–9.

6. One South African study showed up to 40 per cent of trauma patients arrive at hospital in their own vehicle or in other modes of transport. In other parts of the world, it's been estimated this figure might be as high as 90 per cent.

7. A study published in November 2013 showed that clothing presented a greater risk of indirect fracture and a greater severity of those fractures produced. It also suggested that clothing increases infection rates, as the clothing is drawn into the wound, 'acting as a nidus for infection'; http://www.josr-online.com/content/pdf/1749-799X-8-42.pdf

8. http://www.scielo.br/scielo.php?pid=S0100-72032013000900008&script=sci_arttext

9. http://www.sciencedirect.com/science/article/pii/S1072751509016184

10. http://aje.oxfordjournals.org/content/159/7/683.full

11. http://www.irinnews.org/printreport.aspx?reportid=79241

12. http://www.ncbi.nlm.nih.gov/pmc/articles/PMC1570575/

13. 'La Méthode de Traicter les playes Faictes par Harquebutes et Aultres Bastons de Feu'. Paré also developed ways to tie off veins and arteries, making thigh amputations possible, and was one of the first to note how maggots could be used to clean wounds.

14. https://www.sciencenews.org/article/florence-nightingale-passionate-statistician

15. http://www.historynet.com/minie-ball; http://opinionator.blogs.nytimes.com/2012/08/31/the-bullet-that-changed-history/?_r=0

16. https://www.armyheritage.org/education-and-programs/educational-resources/education-materials-index/50-information/soldier-stories/290-civilwarmedicine

17. http://afids.org/publications/PDF/CRI/Prevention%20and%20Management%20of%20CRI%20-4-%20-%20History.pdf

18. http://www.sciencemuseum.org.uk/broughttolife/people/josephlister.aspx

19. Vincent J. Cirillo, *Bullets and Bacilli: The Spanish-American War and Military Medicine* (New Brunswick: Rutgers University Press, 2004), p. 30.

20. R. M. Hardaway, 'Wound Shock: A History of Its Study and Treatment by Military surgeons', *Mil. Med.*, 169, 4, 2004, cited in http://afids.org/publications/PDF/CRI/Prevention%20and%20Management%20of%20CRI%20-4-%20-%20History.pdf/view

21. http://afids.org/publications/PDF/CRI/Prevention%20and%20Management%20of%20CRI%20-4-%20-%20History.pdf

22. http://www.bbc.co.uk/guides/zs3wpv4

23. http://www.ncbi.nlm.nih.gov/pmc/articles/PMC1570575/

24. http://www.sciencemuseum.org.uk/broughttolife/themes/war.aspx

25. http://www.standard.co.uk/news/health/st-marys-adopts-army-tactics-to-save-lives-of-gun-and-knife-victims-8893928.html

26. http://www.thelancet.com/crash-2

27. http://www.popsci.com/article/technology/how-simple-new-invention-seals-gunshot-wound-15-seconds

28. http://www.newscientist.com/article/mg22129623.000-gunshot-victims-to-be-suspended-between-life-and-death.html

29. http://journals.lww.com/jtrauma/Abstract/2014/01000/Unrelenting_ violence___An_analysis_of_6,322.2.aspx; a total of 6,322 patients were treated and inpatient costs were put at $115 million.

30. http://journals.lww.com/annalsplasticsurgery/Abstract/2012/04000/Outcomes_ of_Complex_Gunshot_Wounds_to_the_Hand_and.11.aspx

31. http://journals.lww.com/annalsplasticsurgery/Abstract/2012/04000/Gunshot_ Wounds_to_the_Face__Level_I_Urban_Trauma.12.aspx

32. http://www.ncbi.nlm.nih.gov/pubmed/12616051

33. http://www.bloomberg.com/news/2012-12-21/shootings-costing-u-s-174-billion-show-burden-of-gun-violence.html; http://www.pire.org/documents/ GSWcost2010.pdf

34. http://www.forbes.com/sites/abrambrown/2013/01/14/how-guns-and-violence-cost-every-american-564-in-2010/

35. World Health Organization and World Bank, World Report on Disability (Geneva: WHO, 2011). A 2005 survey revealed that fewer than half of all countries had gun violence rehabilitation programmes.

36. R. L. Leavitt (ed.), *Cross-cultural Rehabilitation: An International Perspective* (London: W. B. Saunders, 1999), p. 99.

37. World Health Organization, Community Based Rehabilitation: CBR Guidelines (Geneva: WHO, 2010).

38. http://www.ippnw.org/pdf/research-kenya-who-pays-price.pdf

39. http://www.uphs.upenn.edu/ficap/resourcebook/pdf/monograph.pdf

Chapter 4: The Suicidal

1. http://www.dailymail.co.uk/news/article-2743457/WHO-calls-action-reduce-global-suicide-rate-800-000-year.html

2. http://www.who.int/mental_health/prevention/suicide/suicideprevent/en/; http://www.bbc.co.uk/news/health-29060238; http://www.cdc.gov/violence prevention/pdf/suicide_datasheet-a.pdf

3. http://www.pewresearch.org/fact-tank/2013/05/24/suicides-account-for-most-gun-deaths/

4. Across the US, in 2010, the gun suicide rate was 6.3 per 100,000 people, compared with 3.6 per 100,000 for gun homicides: http://webappa.cdc.gov/ sasweb/ncipc/dataRestriction_inj.html

5. According to the WISQARS Injury Mortality Report, Alaska has a rate of 15.34 per 100,000 population, New Jersey has a rate of 1.94 per 100,000 population.

6. http://www.ncbi.nlm.nih.gov/pubmed/3553611

7. http://www.cdc.gov/nchs/data/nvsr/nvsr63/nvsr63_03.pdf

8. In 2011, 93 people shot themselves in the UK (out of 6,045 suicides). That

year 19,766 shot themselves in US (out of 38,285): http://www.gunpolicy
.org/firearms/region/united-kingdom and http://www.gunpolicy.org/firearms/
region/united-states

9. https://aoav.org.uk/2014/homicides-in-central-america-up-99-per-cent/
10. http://www.injepijournal.com/content/1/1/6/abstract
11. http://www.smallarmssurvey.org/fileadmin/docs/H-Research_Notes/
    SAS-Research-Note-9.pdf
12. E. G. Richardson and D. Hemenway, 'Homicide, Suicide, and Unintentional
    Firearm Fatality: Comparing the United States with Other High-income
    Countries, 2003', *Journal of Trauma*, 70, 2011, pp. 238–43.
13. http://www.ncbi.nlm.nih.gov/pubmed/7391258
14. C. E. Rhyne, D. I. Templer, L. G. Brown and N. B. Peters, 'Dimensions of
    Suicide: Perceptions of Lethality, Time and Agony', *Suicide and Life-Threatening
    Behavior*, 25, 3, 1995; cited in http://lostallhope.com/suicide-methods/statis-
    tics-most-lethal-methods
15. My research was helped enormously by Scott Anderson's excellent *New
    York Times* report on this: http://www.nytimes.com/2008/07/06/magazine
    /06suicide-t.html?pagewanted=all&_r=0
16. The addition of suicide barrier to a bridge in Washington, DC lowered not
    just the number of suicides that occurred on that bridge, but also the overall
    suicide rate (meaning those people didn't just go and find another bridge to
    jump from).
17. A study published in 2007 by Small Arms Survey estimated that there were
    3.4 million firearms in private households across the country. The Defence
    and Sport Ministry, on the other hand, the same year put the figure at 2.2
    million. Of this number, 535,000 were army weapons, either in the possession
    of current or retired soldiers, or hired out to gun clubs.
18. Of all Swiss men that kill themselves, about one in three shoot themselves.
    Women are more likely to choose less certain methods – such as poisoning,
    or using a sharp implement – while men choose the more lethal courses of
    action.
19. http://www.cato.org/publications/commentary/gun-control-myths-realities
20. http://www.law.harvard.edu/students/orgs/jlpp/Vol30_No2_KatesMauser
    online.pdf
21. http://www.ncbi.nlm.nih.gov/pubmed/23897090
22. http://www.ncbi.nlm.nih.gov/pubmed/21034205
23. http://andrewleigh.org/pdf/GunBuyback_Panel.pdf; even the US Military
    Suicide Research Consortium has found that 'studies demonstrate that method
    substitution is rare'. And that 'the majority of individuals (close to 95 per
    cent) who attempt suicide but are prevented from using their preferred method
    do not die by suicide'.
24. Trends seen in Austria, Brazil, Canada, New Zealand and the UK, to name

some. One survey concluded: 'male firearm suicide rates declined following the introduction of restrictive firearms regulations in Canada'. http://injury-prevention.bmj.com/content/16/4/247.short?g=w_ip_gun_sidetab

25. http://www.nraila.org/news-issues/fact-sheets/1999/suicide-and-firearms.aspx

26. http://www.ncbi.nlm.nih.gov/pmc/articles/PMC1489848/#__ffn_sectitle

27. http://www.nytimes.com/books/99/07/04/specials/hemingway-obit.html

28. http://www.bbc.co.uk/news/magazine-14374296

Chapter 5: The Killers

1. *Report of the Investigation Commission: Kauhajoki School Shooting*, 23 September 2008, p. 44.

2. There is no international, standard definition of 'mass shooting'. In the US, the FBI's definition of mass murder is often used as a starting point: a 'number of murders (four or more) occurring during the same incident, with no distinctive time period between the murders'. This is different from serial killing, where there is 'a temporal separation between the different murders'. The US Congressional Research Service adds more criteria: 'the gunmen do not pursue criminal profit or kill in the name of terrorist ideologies'. For this reason, it excludes religiously motivated attacks.

3. It was something seen in the British media's response to the Northern Ireland Troubles. In the first rank – getting the most prominent coverage – were British people killed in Britain; in the second, the security forces, whether army or RUC; in the third, civilian victims of republicans; and, in the fourth, garnering very little media coverage, were the victims of loyalists – credit to the *Guardian*'s Roy Greenslade for this observation.

4. http://www.fbi.gov/news/stories/2014/september/fbi-releases-study-on-active-shooter-incidents/pdfs/a-study-of-active-shooter-incidents-in-the-u.s.-between-2000-and-2013; the study, of 160 'active shooter incidents' found 46 per cent were in centres of commerce, 24 per cent in centres of learning.

5. Saari was friends with Pekka-Eric Auvinen, a misfit of an eighteen-year-old who had carried out a similar shooting at a school in the Finnish town of Jokela the year before. The pair had played together on an online war game called *Battlefield 2* in which they detonated bombs, shot people and used headsets to communicate. They also sent messages to each other discussing their plans for a shooting. One of the messages said: 'Let's do it together.' They even bought their guns at the same weapons store, a few hundred yards from the school where Auvinen killed eight people and then himself.

6. L. S. De Camp, *The Ancient Engineers* (New York: Ballantine Books,1963), p. 91.

7. http://link.springer.com/article/10.1023%2FA%3A1009691903261; a 2003

study led by Columbia University also found 'ample evidence' of the media's coverage of suicides resulting in more suicides. And a 2011 study in the journal *BMC Public Health* found, unsurprisingly, this effect is especially strong for novel forms of suicide that receive outsize attention in the press.

8. http://www.cbc.ca/news/world/u-s-networks-to-limit-use-of-virginia-tech-killer-video-1.662459

9. http://online.wsj.com/news/articles/SB10001424052702303309504579181702252120052

10. http://www.nytimes.com/1982/04/28/world/seoul-is-stunned-by-policeman-s-slaying-of-56.html

11. http://www.wsj.com/articles/SB10001424053111903554904576464193192352636

12. As with so much in the American world of gun control, the numbers of mass shootings are endlessly debated: the Congressional Research Service say there have only been seventy-eight public mass shootings in the US since 1983. They estimate that over the last three decades, mass shootings have claimed 547 lives and caused 476 injured victims; http://journalistsresource.org/wp-content/uploads/2013/03/MassShootings_CongResServ.pdf

13. http://www.reddit.com/r/GunsAreCool/wiki/2013massshootings

14. Some argue that while mass shootings rose between the 1960s and the 1990s, they actually dropped in the 2000s – that mass killings reached their peak in 1929. There were, accordingly, thirty-two mass shootings in the 1980s, forty-two in the 1990s and twenty-six in the first decade of the century. The challenge is setting the criteria.

15. http://www.washingtonpost.com/wp-srv/special/nation/deadliest-us-shootings/

16. http://www.secretservice.gov/ntac_ssi.shtml

17. http://usatoday30.usatoday.com/news/nation/2006-01-31-postal-shooting_x.htm

18. http://time.com/114128/elliott-rodgers-ucsb-santa-barbara-shooter/

19. http://www.nbcnews.com/id/8917466/ns/nbc_nightly_news_with_brian_williams/t/jonesboro-school-shooter-free-after-seven-years/#.UylZ5_TV9gM

20. http://www.fas.org/sgp/crs/misc/R43004.pdf; the website Mother Jones claimed the average age of a mass shooter is thirty-five.

21. http://forensis.org/PDF/published/2001_OffenderandOffe.pdf; it's also interesting that 'only 6% were judged to have been psychotic at the time of the mass murder'.

22. http://kildall.apana.org.au/autism/articles/bryant.html

23. http://www.cbsnews.com/news/virginia-tech-gunman-warning-signs/

24. http://www.salon.com/2011/01/12/jared_loughner_mass_murderers_diagnose/

25. http://www.npr.org/2012/12/14/167287373/many-mass-killers-have-had-chronic-depression

26. http://edition.cnn.com/2007/US/04/18/vtech.shooting/

27. http://online.wsj.com/news/articles/SB100014240527023033095045791817 02252120052

28. http://www.motherjones.com/politics/2013/02/assault-weapons-high-capacity -magazines-mass-shootings-feinstein

29. http://edition.cnn.com/2012/07/23/opinion/webster-aurora-shooter/

30. http://www.washingtonpost.com/blogs/wonkblog/files/2013/02/mass_shoot ings_2009-13_-_jan_29_12pm1.pdf; almost all weapons in mass shootings are legally purchased – as Mother Jones concluded: 'of the 143 guns possessed by the killers, more than three quarters were obtained legally'.

31. Breivik had saved €2,000 to spend on a 'high-class' prostitute before his planned massacre.

32. http://www.vice.com/en_uk/read/utoya-ptsd-professor-lars-weisth-anders-breiv ik-341

33. A licence is required to own a gun, and the owner must give a statement as to why they want one. Many categories of guns, including automatics and some handguns, are banned from sale altogether.

34. A ban on semi-automatic rifles was introduced in September 2011, but it was lifted at the end of February 2013.

35. http://www.gq.com/news-politics/newsmakers/201311/fake-hitman-murder-for-hire

36. http://www.forbes.com/sites/andygreenberg/2013/11/21/alleged-silk-road-ross -ulbricht-creator-now-accused-of-six-murder-for-hires-denied-bail/

37. http://www.washingtonpost.com/blogs/the-switch/wp/2013/10/02/silk-roads-mastermind-allegedly-paid-80000-for-a-hitman-the-hitman-was-a-cop/

38. The assassination of Abraham Lincoln was not the original plan. John Wilkes Booth and a group of co-conspirators were meant to kidnap the president and hold him hostage in exchange for prisoners. But the plan quickly changed to assassinating Lincoln, as well as the vice president and secretary of state.

39. As he was shot in the neck, this seems unlikely: http://qi.com/infocloud/ the-first-world-war

40. http://www.theguardian.com/film/2012/feb/17/woody-harrelson-my-father-contract-killer-rampart

41. http://www.nytimes.com/1982/11/30/us/murder-trial-defendant-says-us-is-hiding-facts.html; he also claimed to have been involved in the assassination of John F. Kennedy: http://www.theguardian.com/film/2012/feb/17/woody-harrelson-my-father-contract-killer-rampart

42. http://edition.cnn.com/2013/03/01/world/americas/mexico-young-assassin/

43. http://news.bbc.co.uk/1/hi/world/europe/4801971.stm

44. http://www.theguardian.com/world/2011/aug/04/indian-police-capture-alleged -contract-killer

45. http://www.ndtv.com/article/cities/hired-gun-killer-of-100-people-shot-dead-in-ghaziabad-307025

46.   He thought he was going to get £2,000: http://www.theguardian.com/world/2011/may/24/britains-youngest-hitman-jailed-life

47.   http://onlinelibrary.wiley.com/doi/10.1111/hojo.12063/abstract;jsessionid=271A531DBBB369D8C925ED565173A509.f02t03; http://www.theguardian.com/uk-news/2014/jan/25/hitmen-for-hire-secrets-contract-killers

Chapter 6: The Criminals

1.   http://www.smallarmssurvey.org/fileadmin/docs/A-Yearbook/2010/en/Small-Arms-Survey-2010-Chapter-04-EN.pdf

2.   The striking thing, though, when I began my research into gangs, was how little journalists and writers had focused on the guns wielded by criminals. Their rituals, clothing or nicknames were the things that turned gangland lives into the stuff of black legend. But the gangs' guns seemed often just a sideline fact.

3.   It is young men, those aged between fifteen and twenty-nine, who are not only the most likely to be in gangs, but who are also killed by guns. They account for half of all global firearm homicide victims, with up to 100,000 deaths every year. http://www.smallarmssurvey.org/fileadmin/docs/A-Yearbook/2006/en/Small-Arms-Survey-2006-Chapter-12-EN.pdf

4.   http://www.insightcrime.org/news-analysis/latin-america-worlds-most-violent-region-un

5.   http://abcnews.go.com/Blotter/story?id=4695848

6.   United Nations Office on Drugs and Crime, 2011 *Global Study on Homicide: Trends, Context, Data* (Vienna: UNODC, 2011), pp. 93, 114.

7.   http://www.insightcrime.org/news-analysis/is-el-salvador-negotiating-with-street-gangs

8.   Given how many murders are often 'hidden' in this part of the world, I have taken the upper limits quoted in the press here. Other reports have said the homicide rate dropped from fourteen murders a day to about five: http://www.wola.org/commentary/one_year_into_the_gang_truce_in_el_salvador This number has been repeated, including by the BBC: http://www.bbc.co.uk/news/world-latin-america-18517208 But this report does put the pre-truce figures as '11 to 17 murders per day in 2011': http://www.cipamericas.org/archives/11591

9.   https://aoav.org.uk/wp-content/uploads/2014/11/the_devils_trade_lr.pdf; firearms are very visible in El Salvador. Every gas station, shopping mall or even hole-in-the-wall store seems to have security guards armed with revolvers and 12-gauge shotguns. With some 85,000 nationwide, security guards form the largest armed force in the country, almost three times the police and the army combined. And they hold the bulk of the estimated 450,000 guns legally owned.

10.   https://aoav.org.uk/wp-content/uploads/2014/11/the_devils_trade_lr.pdf; 2011

weapons confiscation data from the Ministry of Justice shows that, of a total of 4,097 firearms confiscated in the first eleven months of that year, 78 per cent, or 3,208 weapons, were either revolvers or semi-automatic pistols. Shotguns (433) and rifles (235) were much less common. Finally, explicitly military weapons such as grenades (58), carbines (27) and light machine-guns (29) were confiscated in extremely small numbers, which presumably reflects their relative rarity along with their greater material and financial value.

11. http://www.cpdsindia.org/smallarmsresearchfiles1.htm; the Salvadoran government claims it destroyed some 28,036 weapons between 2006 and 2008, but this is very hard to corroborate. The military's transparency about its holdings of weapons is sharply lacking.

12. http://www.ourworldindata.org/data/violence-rights/ethnographic-and-archaeo logical-evidence-on-violent-deaths; an Oxfam study on violence in PNG's Highlands reported that in 80 per cent of traumatic injuries a weapon had been used. Oxfam Position Paper: 'Armed Violence and the Links to Human Security in Papua New Guinea', http://www.oxfam.org.nz/report/oxfam-position-paper-armed-violence-and-the-links-to-human-security-in-papua-new -guinea

13. Raymond C. Kelly, *Warless Societies and the Origin of War* (Ann Arbor: The University of Michigan Press, 2000), p. 21.

14. Simon Harrison, *Violence, Ritual and the Self in Melanesia* (Manchester: Manchester University Press, 1993), p. 88.

15. P. Alpers, 'Papua New Guinea: Small Numbers, Big Fuss, Real Results', in A. Karp (ed.), *The Politics of Destroying Surplus Small Arms* (London: Routledge, 2009), p. 155.

Chapter 7: The Police

1. Russia has about 1,550,000 police guns and the US about 1,150,000 – with similar ratios per officer as India. http://www.smallarmssurvey.org/fileadmin/docs/H-Research_Notes/SAS-Research-Note-24-Annexe.pdf

2. American state and local police officers, on the other hand, have an average of 1.3 official firearms each – http://www.smallarmssurvey.org/fileadmin/docs/H-Research_Notes/SAS-Research-Note-24.pdf

3. There are an estimated 500,000 to 4 million guns in South Africa: http://africacheck.org/wp-content/uploads/2013/03/The-Proliferation-of-Firearms-in-South-Africa-1994-2004.pdf

    About 5 million people in South Africa do not have access to clean tap water. http://africacheck.org/reports/claim-that-94-of-south-aclaim-that-94-in-sa-have-access-to-safe-drinking-water-doesnt-hold-water/

    The Cape gangs reportedly hold about 10 per cent of the guns in the region:

http://www.smallarmssurvey.org/fileadmin/docs/A-Yearbook/2010/en/Small-Arms-Survey-2010-Chapter-04-EN.pdf

4. N. M. Campbell, J. G. Colville, Y. van der Heyde and A. B. van As, 'Firearm Injuries to Children in Cape Town, South Africa: Impact of the 2004 Firearms Control Act', *The South African Journal of Surgery (SAJS)* 51, 3, 2013, p. 92; http://www.theguardian.com/world/2013/feb/26/south-african-guns-are-us

5. http://www.theguardian.com/world/2014/may/29/gangs-south-africa-western-cape

6. Coloured is the ethnic label for those of mixed ethnic origin – their ancestry from Europe, Asia, and the various Khoisan and Bantu tribes of southern Africa.

7. Figure quoted by Andre Standing in a 2005 Institute of Security Studies policy discussion paper.

8. http://www.iol.co.za/news/crime-courts/bonteheuwel-residents-say-cops-are-failing-1.1683639#.U-s6-Vbobx4

9. http://www.issafrica.org/uploads/LEGGETT2.PDF

10. http://www.news24.com/SouthAfrica/News/Cape-Town-cop-shot-dead-20130723

11. The gangs often tried to get their hands on police pistols; in 2009, nearly 3,000 police guns were lost or taken; http://www.telegraph.co.uk/news/worldnews/africaandindianocean/southafrica/7085320/South-Africa-police-lose-3000-guns-a-year.html

12. http://www.issafrica.org/uploads/LEGGETT2.PDF

13. It's estimated there were 4,600 deaths as a result of police action in the seven years after apartheid ended; D. Bruce, 'Interpreting the Body Count: South African Statistics on Lethal Police Violence', *South African Review of Sociology*, 36, 2, 2005, pp. 141–59.

14. Of the shots fired 34 per cent were in barricade scenarios, 36 per cent involved hostages and 21 per cent a suicidal subject – death by cop. First published in *Tactical Response*, September/October 2005; http://www.hendonpub.com/resources/article_archive/results/details?id=3879

15. http://rt.com/usa/police-sniper-suicidal-boy-870/

16. http://www.policestateusa.com/2014/michael-blair/; see also: http://www.dailymail.co.uk/news/article-2535424/Fury-schizophrenic-teen-Keith-Vidal-shot-dead-Southport-Police.html and http://articles.latimes.com/2000/nov/01/news/mn-45199

17. http://www.policestateusa.com/2014/iyanna-davis/

18. Ibid.

19. http://longisland.newsday.com/templates/simpleDB/?pid=345&currentRecord=1501

20. In the thirty-five police 'deadly force incidents' since 2001, a firearm was only recovered from the suspected criminal in twelve cases. Nassau police officers had also shot at someone in a moving car at least ten times since 2006, despite

department protocols forbidding firing at a moving vehicle; credit to these observations must go to http://data.newsday.com/long-island/data/crime-and-punishment/nassau-deadly-force/; the Nassau Police did not respond to my request for an interview.

21. http://www.fbi.gov/about-us/cirg/tactical-operations; in 1987 the Minneapolis Police Department conducted thirty-six SWAT team raids. In 1996 the same department carried out 700 raids. In 1989 the number of police officers in the tactical operations branch of the Portland, Oregon, Police Department was two. By 1994 this number was fifty-six. This was even as violent crime fell.

22. http://nineronline.com/2011/10/unc-charlotte-swat-team---an-asset-we-hope-to-never-use/

23. http://online.wsj.com/news/articles/SB10001424127887323848804578608040780519904

24. http://www.globalresearch.ca/the-militarization-of-police-state-usa/5377240?print=1

25. http://edition.cnn.com/2009/CRIME/11/24/georgia.gay.club.lawsuit/

26. http://articles.orlandosentinel.com/2010-11-07/health/os-illegal-barbering-arrests-20101107_1_criminal-barbering-licensing-inspections-dave-ogden

27. http://www.motherjones.com/politics/2010/11/aiyana-stanley-jones-detroit

28. http://www.unionleader.com/article/20140919/NEWS03/140918916

29. http://www.policestateusa.com/2014/swat-throws-grenade-in-playpen/

30. In 2014, fifty US police officers were shot and killed in the line of duty. In the fifty years before 2014, thirty-one police officers were unlawfully shot and killed in the line of duty in England, Wales and Scotland. Thanks to the Police Roll of Honour for this information: http://www.policememorial.org.uk

31. http://www.theguardian.com/commentisfree/2013/oct/07/militarization-local-police-america

32. As *The Economist* reported: 'In 1986, its first year of operation, the federal Assets Forfeiture Fund held $93.7m. By 2012, that and the related Seized Assets Deposit Fund held nearly $6 billion'; http://www.economist.com/news/united-states/21599349-americas-police-have-become-too-militarised-cops-or-soldiers

33. http://www.denverpost.com/nationworld/ci_22594279/homeland-security-aims-buy-1-6-billion-rounds

34. The company also found that in the summer months, gunfire increases, and that 42 per cent of all gunfire happened in June, July and August. In the worst place that they looked at, they found on average over eight bullets were shot every single day for an entire year within a single square mile; http://www.shotspotter.com/policy-implications

35. Others would contest this figure as being on the low side. As the *Washington Post* reported: 'Officials with the Justice Department keep no comprehensive

database or record of police shootings, instead allowing the nation's more than 17,000 law enforcement agencies to self-report officer-involved shootings as part of the FBI's annual data on "justifiable homicides" by law enforcement'; http://www.washingtonpost.com/news/post-nation/wp/2014/09/08/how-many-police-shootings-a-year-no-one-knows/; the website http://www.fatal encounters.org has an arguably more accurate figure and these numbers are quoted later in this book.

36. http://www.economist.com/blogs/democracyinamerica/2014/08/armed-police; http://www.smallarmssurvey.org/fileadmin/docs/A-Yearbook/2011/en/Small-Arms-Survey-2011-Chapter-03-EN.pdf

37. https://www.ncjrs.gov/pdffiles1/nij/grants/204431.pdf

38. http://billmoyers.com/2014/08/13/not-just-ferguson-11-eye-opening-facts-about-americas-militarized-police-forces/

39. http://www.economist.com/news/united-states/21591877-when-pupils-get-trouble-silly-reasons-results-can-be-serious-perils

40. http://www.nytimes.com/2013/01/17/education/report-criticizes-school-discipline-measures-used-in-mississippi.html?_r=0

41. http://opinionator.blogs.nytimes.com/2014/02/15/one-nation-under-guard/

42. http://www.hrw.org/sites/default/files/reports/philippines0409webwcover_0.pdf

43. http://www.hrw.org/sites/default/files/reports/philippines0514_ForUpload_0_0_1.pdf

44. http://www.hrw.org/ru/node/82034/section/12

45. http://content.time.com/time/magazine/article/0,9171,501020701-265480,00.html

46. http://www.amnesty.org/en/news/brazil-police-still-have-blood-their-hands-20-years-massacre-2013-07-24; police in the United States arrest 37,000 people for every person they kill.

47. http://www.economist.com/blogs/americasview/2014/03/police-violence-brazil; the US Department of State's annual report for Brazil in 2013 states that: 'human rights problems included excessive force and unlawful killings by state police; excessive force, beatings, abuse, and torture of detainees and inmates by police and prison security forces'.

48. http://www.smallarmssurvey.org/about-us/highlights/highlight-iava-ib3.html; from 791 in 2004 to 1,421 in 2010.

49. http://www.davekopel.com/2A/Foreign/Brown-Journal-Kopel.pdf

Chapter 8: The Military

1. Many of these twenty-one countries are small tropical islands in the Caribbean or the South Pacific, like Saint Lucia or Vanuatu. Their defence

is, largely, their remoteness or their natural environment; both deterring an invasion.

2. http://www.smallarmssurvey.org/fileadmin/docs/H-Research_Notes/SAS-Research-Note-34.pdf; in South America, for instance, it's thought about 1.3 million military guns are 'undoubtedly superfluous'; Argentina has over a half million guns they don't need – over 77 per cent of its total number – and in Guyana as many as 83 per cent of their military guns may be surplus to requirement.

3. L. Themner and P. Wallensteen, 'Armed Conflicts, 1946–2010', *Journal of Peace Research*, 48, 4, 2011, pp. 525–36.

4. http://ploughshares.ca/pl_publications/the-wars-of-1997-introduction-to-the-armed-conflicts-report-1998/

5. http://www.genevadeclaration.org/measurability/global-burden-of-armed-violence/global-burden-of-armed-violence-2011.html; *Global Burden of Armed Violence* (2011). This figure of 55,000 conflict deaths a year seems to accord with another major analysis of deaths in wars that happened between 1946 and 2002 – as outlined in *The Human Security Report*, published in 2005. That review concluded there has been a clear but even decline in battle deaths since the Second World War and that wars between states were deadlier than civil wars.

6. Joakim Kreutz and Nicholas Marsh, 'Lethal Instruments: Small Arms and Deaths in Armed Conflict', *Small Arms, Crime and Conflict*, 49, 2011.

7. Ibid.

8. http://www.dtic.mil/dtic/tr/fulltext/u2/a498077.pdf#page=68; http://journals.lww.com/jtrauma/Abstract/2004/08000/A_U_S__Army_Forward_Surgical_Team_s_Experience_in.1.aspx

9. Made by F. N. Herstal, 200,000 of these general-purpose machine-guns are in use by over ninety militaries around the world.

10. http://www.smallarmssurvey.org/fileadmin/docs/F-Working-papers/SAS-WP1-Iraq.pdf

11. http://www.washingtonpost.com/wp-dyn/content/article/2007/11/16/AR2007111600865.html; http://www.gao.gov/new.items/d05687.pdf

12. http://www.independent.co.uk/news/world/americas/us-forced-to-import-bullets-from-israel-as-troops-use-250000-for-every-rebel-killed-314944.html

13. http://www.gao.gov/new.items/d05687.pdf

14. Ibid.

15. http://www.bbc.co.uk/news/world-middle-east-11107739

16. http://www.fas.org/sgp/crs/natsec/RL33110.pdf

17. https://www.iraqbodycount.org (accessed on 21 April 2014). This reached a feverish peak in 2006 with 2,334 gunfire-related deaths in July 2006 and 2,118 in October 2006.

18. And the numbers of civilians dying at checkpoints got worse as the war went

on. In 2004, when I was there, the war logs showed twenty-two civilian deaths. By 2005 it was nearly 300.

19. http://www.washingtonpost.com/pb/world/haditha-iraq-haunted-by-marines-shooting-spree/2011/12/09/gIQAEzJblO_gallery.html#item0

20. http://articles.latimes.com/2014/feb/18/nation/la-na-nn-soldier-suicide-rape-iraq-girl-mahmoudiya-20140218

21. http://www.ipsnews.net/2012/11/israel-ranked-as-worlds-most-militarised-nation/

22. http://www.gunpolicy.org/firearms/region/israel; Aaron Karp, 'Trickle and Torrent: State Stockpiles', *Small Arms Survey 2006: Unfinished Business* (Oxford: Oxford University Press, 2006), Chapter 2 (Appendix I), p. 61.

23. www.gunpolicy.org

24. http://www.idfblog.com/blog/2013/12/01/eyes-like-hawk-snipers-discover-secrets-special-training/

25. One British soldier, among the 330 trained snipers reported to be operating in the UK forces, was said to have killed thirty-nine Taliban. Each of the 8.59mm bullets used by UK snipers in southern Afghanistan had cost about £20, compared to the Javelin anti-tank missile, which then cost £70,000.

26. http://www.btselem.org/statistics/fatalities/any/by-date-of-death/wb-gaza/pal estinians-killed-during-the-course-of-a-targeted-killing

27. http://www.theguardian.com/uk/2009/feb/15/army-taliban-sniping; though the sniper's art in war certainly pre-dated this. Faced with the devastating impact of mechanised gun slaughter, General Hiram Berdan from the North and General Robert E. Lee from the South in the American Civil War both set up units of designated sharpshooters.

28. After 550 kills, though, Häyhä was shot in his lower left jaw by a Russian soldier. He did not die, regaining consciousness on 13 March 1939, the day peace was declared. The Finns lost 22,830 men compared to 126,875 Russians, who had an invading force 1.5 million strong. As one Red Army general recalled, 'We gained 22,000 square miles of territory. Just enough to bury our dead.'

29. http://www.dailymail.co.uk/news/article-1270414/British-sniper-sets-new-sharpshooting-record-1-54-mile-double-Taliban-kill.html

30. A. Wacker, *Sniper on the Eastern Front: The Memoirs of Sepp Allerberger Knights Cross* (Pen & Sword, 2005), p. 149.

31. Ibid., p. 119.

32. http://news.sky.com/story/678761/israeli-army-t-shirts-mock-gaza-killings

33. http://www.theguardian.com/world/2012/jul/26/jewish-population-west-bank-up

34. http://www.washingtoninstitute.org/policy-analysis/view/violence-by-extrem ists-in-the-jewish-settler-movement-a-rising-challenge

35. Credit to Max Hastings for these observations: http://www.dailymail.co.uk/

debate/article-2703531/MAX-HASTINGS-Ive-loved-Israel-brutality-breaks-heart.html

36. http://www.btselem.org/statistics/fatalities/after-cast-lead/by-date-of-event/wb-gaza/palestinian-minors-killed-by-israeli-security-forces

37. http://www.unicef.org/sowc96/2csoldrs.htm

38. http://www.warchild.org.uk/issues/child-soldiers

39. Ibid. The use of children carrying guns marks the conscience of modernity's conflicts. Boys as young as ten were used by the Khmer Rouge. Children made up about 50 per cent of the insurgency in Sierra Leone. Joseph Kony's Lord Resistance army in Uganda abducted at least 20,000 children to become fighters or sex slaves. The list of countries that have used children carrying guns in the recent past goes on . . . and on: Burundi, Afghanistan, the Central African Republic, Burma, India, Iran, Iraq, Palestine, Lebanon, Nepal, the Philippines, Sri Lanka, Syria, Thailand, Bolivia, Colombia. Even the US army acknowledged that around sixty seventeen-year-olds were deployed to Iraq and Afghanistan in 2003 and 2004.

Chapter 9: The Civilians

1. http://www.smallarmssurvey.org/fileadmin/docs/H-Research_Notes/SAS-Research-Note-9.pdf

2. http://www.reuters.com/article/2007/08/28/us-world-firearms-idUSL2834893820070828

3. More specifically – 90 guns per 100 people. http://www.smallarmssurvey.org/about-us/highlights/highlight-research-note-9-estimating-civilian-owned-firearms.html

4. http://www.smallarmssurvey.org/fileadmin/docs/H-Research_Notes/SAS-Research-Note-9.pdf

5. Conversation with leading firearm prop industry expert.

6. http://www.airgunshooting.co.uk/expert-advice/airgun_law_in_the_uk_1_1111764

7. A law enacted in 2005 regulating the use of weapons decreed that only the Ministry of the Interior and the Ministry of National Defense were permitted to operate firing ranges in Cambodia.

8. http://www.slate.com/articles/news_and_politics/dispatches/features/2004/dispatches_from_cambodia/gunshopping_in_phnom_penh.html

9. http://www.smallarmssurvey.org/fileadmin/docs/A-Yearbook/2006/en/Small-Arms-Survey-2006-Chapter-05-EN.pdf; http://www.gunpolicy.org/firearms/region/cambodia

10. http://www.fas.org/sgp/crs/misc/RL32842.pdf – this works out at 114 million handguns, 110 million rifles, and 86 million shotguns. The second country

in the world in terms of gun ownership, Yemen, has significantly fewer – about 55 firearms per 100 people. At the other end of the spectrum is Japan, where there is less than one firearm for every 100 people.

11. http://www.ncbi.nlm.nih.gov/pmc/articles/PMC2610545/; putting it simply, it's that guy in Alabama – with a basement filled with fourteen handguns, twelve rifles and nineteen shotguns – that helps make up these huge figures.

12. http://www.theguardian.com/news/datablog/2012/dec/17/how-many-guns-us

13. http://religion.blogs.cnn.com/2013/02/05/arkansas-to-allow-concealed-guns-in-churches/

14. http://www.salon.com/2012/12/18/7_craziest_gun_laws_in_america/

15. In the state of Montana a long gun can be sold to a fourteen-year-old for hunting. http://smartgunlaws.org/minimum-age-to-purchase-possess-firearms-policy-summary/

16. http://talkingpointsmemo.com/dc/missouri-republican-wants-to-make-it-a-felony-for-his-fellow-lawmakers-to-propose-gun-laws

17. http://www.huffingtonpost.com/2013/08/23/nelson-georgia-guns_n_3805292.html

18. The thinking behind this argument is succinctly put forward in this report: http://www.washingtonpost.com/blogs/govbeat/wp/2013/11/08/which-of-the-11-american-nations-do-you-live-in/

19. http://www.nssf.org/PDF/research/TargetShootingInAmericaReport.pdf

20. http://www.yorkshirepost.co.uk/news/features/exposed-deadly-us-cult-of-the-sniper-1-2438333

21. Including homicides, suicides and accidental deaths.

22. http://www.huffingtonpost.com/adam-winkler/did-the-wild-west-have-mo_b_956035.html; modern-day Tombstone has fewer restrictive gun laws than its lawless Old West equivalent.

23. http://www.bbc.co.uk/manchester/2002/events/shooting.shtml

24. http://www.nra.org.uk/common/asp/general/history1.asp?site=NRA

25. Except for St Louis in 1904 and Amsterdam in 1928. http://www.teamgb.com/summer-sports/shooting

26. From 1908 to 1948 Olympic shooters competed in Running Deer events. There competitors fired at moving deer silhouettes from 100 metres away. They scored points by hitting one of three concentric circles in the deer's vital organs. https://www.usashooting.org/library/Olympic/Shooting_History.pdf

27. http://www.scotsman.com/sport/final-curtain-the-last-live-pigeon-shooting-event-at-the-olympic-games-1900-1-1085923

28. http://www.topendsports.com/events/discontinued/shooting-duelling-pistol.htm

29. In the rifle and pistol classes shooters fired at ten-ring targets and in the shotgun events they aimed at clay targets, released on a shooter's command.

30. http://www.olympic.org/shooting-equipment-and-history?tab=history

31. http://www.olympic.org/karoly-takacs

32. Homicide counts and rates, time series 2000–12; http://www.unodc.org/gsh/en/data.html

33. http://www.gunpolicy.org/firearms/region/iceland

34. Of a population of 325,000 Iceland has only 136 people in prison. The US may have a population 1,000 times that of Iceland but it has 2.2 million people in prison; per capita this is about 1,500 times greater than in Iceland.

35. http://www.amren.com/news/2013/05/why-is-violent-crime-so-rare-in-iceland/

36. http://commodityhq.com/commodity/alternatives/antique-guns/

37. http://www.mnh.si.edu/onehundredyears/expeditions/SI-Roosevelt_Expedition.html

Chapter 10: The Hunters

1. http://www.theatlantic.com/magazine/archive/2008/12/timeline-somalia-1991-2008/307190/

2. http://www.telegraph.co.uk/news/worldnews/africaandindianocean/somalia/1444828/British-aid-couple-killed-at-Somali-school.html

3. The Fourth Council of the Lateran of 1215, held under Pope Innocent III, decreed: 'We interdict hunting or hawking to all clerics.'

4. Hunter-gatherers still exist: in the Amazon (Aché); Africa (the San people and the Hadza of Tanzania); New Guinea (the Fayu); Thailand and Laos (the Mlabri); and Sri Lanka (the Vedda); not to mention a handful of uncontacted peoples.

5. http://www.dailymail.co.uk/news/article-2508209/Winchester-Deadly-Passion-presenter-Melissa-Bachman-sparks-outrage-posing-lion-shot-dead.html

6. The possible caveat is that a hunted animal might be shot and injured and not killed outright by a huntsman's bullet. But then again many animals may sense their looming death when they are herded into an abattoir.

7. http://www.economist.com/blogs/graphicdetail/2012/04/daily-chart-17

8. http://www.theguardian.com/tv-and-radio/2011/dec/01/nature-urbanisation-david-attenborough

9. http://www.bbc.co.uk/news/science-environment-26140827

10. http://dianamandache.com/auction-shotgun-king-of-romania/

11. http://www.face.eu/sites/default/files/documents/english/face_annual_report_2013_en.pdf

12. http://www.face.eu/sites/default/files/documents/english/position_paper_hunttour_-_en.pdf

13.   http://www.slate.com/articles/news_and_politics/explainer/2007/11/packing_
      heat_in_helsinki.html

14.   http://www.face.eu/sites/default/files/attachments/data_hunters-region_
      sept_2010.pdf

15.   http://www.conservationforce.org/role4.html;   http://www.nssf.org/PDF/
      research/HuntingInAmerica_EconomicForceForConservation.pdf

16.   http://www.gallup.com/poll/20098/gun-ownership-use-america.aspx

17.   At least this was the case in 2004, and shooting has become more popular
      since then.   http://www.nraila.org/news-issues/fact-sheets/2004/nra-ila-
      hunting-fact-card.aspx

18.   The term appears as early as 1748 in the journal of Conrad Weiser, who
      wrote, while travelling through 'Indian territory' (in what is now Ohio) in
      1748: 'He has been robbed of the value of 300 Bucks.'

19.   http://www.nssf.org/PDF/research/HuntingInAmerica_EconomicForceFor
      Conservation.pdf

20. $37.9 billion (in 2011): http://www.ewebmarketing.com.au/blog/google-
      rakes-in-37-9-billion-in-revenue-for-2011/

21.   http://econpost.com/vermonteconomy/vermont-gdp-size-rank

22.   http://nssf.org/newsroom/releases/show.cfm?PR=081512_USFWS.cfm&path=2012

23.   http://www.ihea.com;   http://www.gunbabygun.com/gun-baby-gun/data-
      shows-hidden-human-cost-hunting-us-80-gun-deaths-year/

24.   http://www.bbc.co.uk/news/business-20204594

25.   http://www.independent.co.uk/news/uk/home-news/im-experiencing-
      austerity-as-well-says-princess-michael-of-kent-8993437.html

26.   http://www.monbiot.com/2014/04/28/the-shooting-party/

27.   http://news.wildlife.org/twp/game-ranching-in-south-africa/

28.   http://www.africanindaba.com/2013/09/some-interesting-facts-about-the-
      hunting-industry-in-south-africa/

29.   http://edition.cnn.com/2014/01/16/us/black-rhino-hunting-permit/

30.   http://www.ifaw.org/sites/default/files/Lions%20Fact%20Sheet.pdf

31.   Others take it one step further and want to bag the Tiny Ten (the damara
      dik-dik, the blue, grey and red duiker, the southern and northern grysbok,
      the klipspringer, oribi, steenbok and suni).

32.   http://news.wildlife.org/twp/game-ranching-in-south-africa/

Chapter 11: The Sex Pistols

1.   http://www.sigsauer.com/CatalogProductList/pistols-mosquito.aspx

2.   The record for most kills achieved by a female sniper is held by Lyudmila M.
      Pavlichenko, a Russian sniper in WWII. She is credited with killing 309
      enemy soldiers.

3. http://www.wired.com/2008/11/return-of-white/; https://www.youtube.com/watch?v=BaeQeGpetvA#t=52

4. http://www.telegraph.co.uk/news/worldnews/northamerica/usa/1411077/A-snipers-life.html

5. http://www.snopes.com/medical/doctor/drruth.asp

6. http://www.gallup.com/poll/160223/men-married-southerners-likely-gun-owners.aspx ; http://www.people-press.org/2013/03/12/section-3-gun-owner-ship-trends-and-demographics/

7. Data from the National Department of Arms and Explosives on applications for firearms licences, 2008–09; http://www.smallarmssurvey.org/fileadmin/docs/A-Yearbook/2013/en/Small-Arms-Survey-2013-Chapter-2-EN.pdf

8. http://www.theguardian.com/money/2013/dec/18/what-are-the-best-paid-jobs-uk-2013

9. Lawrence Mishel, Josh Bivens, Elise Gould, Heidi Shierholz, *The State of Working America*, 12th edition (Ithaca, NY: Cornell University Press, 2012).

10. http://www.aauw.org/research/the-simple-truth-about-the-gender-pay-gap/

11. http://youliveyourlife.com

12. http://historynewsnetwork.org/article/154225

13. http://store.waltherarms.com/very-tough-ppx-shirt.html

14. http://www.salon.com/2012/12/17/bushmasters_horrible_ad_campaign/

15. Small Arms Survey, *Unfinished Business* (Oxford: Oxford University Press, 2006), p. 317; http://www.smallarmssurvey.org/fileadmin/docs/A-Yearbook/2006/en/Small-Arms-Survey-2006-Chapter-12-EN.pdf

16. http://kdvr.com/2012/12/28/colorado-columnist-assault-rifle-owners-have-tiny-penises/

17. http://www.bloombergview.com/articles/2013-03-08/how-brazil-exploited-sexual-insecurity-to-curb-guns-an-interview-with-antonio-bandeira

18. Stephen Marche has written a very good article on this and must take credit for these thoughts: http://www.esquire.com/features/thousand-words-on-culture/guns-are-beautiful-0313

19. http://www.ncbi.nlm.nih.gov/pubmed/16866740; http://www.nytimes.com/2006/05/09/health/09guns.html?_r=0

20. https://dspace.lboro.ac.uk/dspace-jspui/handle/2134/5932

21. The ratios are 9.7 versus 2.7 per 100,000. It is the highest in the Americas (29.3 per 100,000 males), where it is nearly seven times higher than in Asia, Europe and Oceania (all under 4.5 per 100,000 males); https://www.unodc.org/documents/data-and-analysis/statistics/GSH2013/2014_GLOBAL_HOMICIDE_BOOK_web.pdf, p. 13

22. http://www.huffingtonpost.com/robert-muggah/how-to-end-brazils-homici_b_4556945.html

23. http://www.who.int/violence_injury_prevention/violence/world_report/en

24. Owen Greene and Nicholas Marsh (eds.), *Small Arms, Crime and Conflict: Global Governance and the Threat of Armed Violence* (London: Routledge, 2011), p. 48.

25. http://www.who.int/bulletin/volumes/86/9/07-043489/en/. One review did find out that, in China, more women killed themselves than men: http://www.ncbi.nlm.nih.gov/pmc/articles/PMC1730718/pdf/v007p00104.pdf

26. As of January 2015, if you do this, please make sure you place these phrases between quotation marks to search for the exact text on Google.

27. http://www.telegraph.co.uk/news/worldnews/centralamericaandthecaribbean/mexico/6409484/Mexican-city-is-murder-capital-of-the-world.html; http://smallwarsjournal.com/jrnl/art/the-mexican-undead-toward-a-new-history-of-the-%E2%80%9Cdrug-war%E2%80%9D-killing-fields

28. In 2012 in Latin American and the Caribbean, 91 per cent of homicide victims (101,041 cases) were men, and 9 per cent women (9,704 cases): https://aoav.org.uk/2014/homicides-in-central-america-up-99-per-cent/

29. http://papers.ssrn.com/sol3/papers.cfm?abstract_id=1112308; http://texas-center.tamiu.edu/PDF_BR/V7/v7-Albuquerque.pdf

30. Between 1990 and 2005.

31. http://www.texasobserver.org/qa-molly-molloy-story-juarez-femicides-myth/

32. Based on a 2011 homicide rate of 7.8 per 100,000 in Pakistan, 5.1 per 100,000 in the USA, and 1.03 in the UK. United Nations Office on Drugs and Crime, *Homicide in 207 Countries. Global Study on Homicide 2011: Trends, Context, Data; Statistical Annex* (Vienna: UNODC, 2011) and historical population data, United States Census Bureau (USCB) International Data Base (Suitland, MD: US Census Bureau Population Division).

33. http://www.memri.org/report/en/0/0/0/0/0/0/6484.htm

34. http://blogs.independent.co.uk/2012/01/16/it's-a-girl-the-three-deadliest-words-in-the-world/

35. http://edition.cnn.com/2014/07/29/opinion/giffords-gun-violence-congress/; in the US, firearm assaults on female family members, and intimate acquaintances are approximately twelve times more likely to result in death than assaults using other weapons.

36. http://www.huffingtonpost.com/2014/06/18/guns-domestic-abuse_n_5506643.html

37. http://cdn.americanprogress.org/wp-content/uploads/2014/06/GunsDomesticViolence2.pdf

38. He made this remark in a speech at the Conservative Political Action Committee: http://www.salon.com/2013/03/15/at_cpac_wayne_lapierre_tackles_rape/

39. http://annals.org/article.aspx?articleid=1814426#r4-6

40. http://www.smallarmssurvey.org/fileadmin/docs/A-Yearbook/2013/en/Small-Arms-Survey-2013-Chapter-2-EN.pdf

41. Ibid.

42. http://www.ncbi.nlm.nih.gov/pubmed/11991417

43. http://mediamatters.org/blog/2014/04/18/guns-make-domestic-violence-deadlier/198942

44. http://ajl.sagepub.com/content/early/2011/02/01/1559827610396294.abstract

45. http://www.telegraph.co.uk/women/womens-life/9969670/Do-guns-make-women-any-safer.html; http://www.fbi.gov/about-us/cjis/ucr/crime-in-the-u.s/2012/crime-in-the-u.s.-2012/tables/5tabledatadecpdf/table_5_crime_in_the_united_states_by_state_2012.xls

46. D.J. Wiebe, 'Homicide and Suicide Risks Associated with Firearms in the Home: A National Case-Control Study', *Annals of Emergency Medicine*, Volume 41, January–June 2003.

47. France Winddance Twine, *Girls With Guns: Firearms, Feminism, and Militarism* (London: Routledge, 2013), p. 8.

48. http://www.gallup.com/poll/150353/self-reported-gun-ownership-highest-1993.aspx

49. http://www.nytimes.com/2011/05/24/us/24crime.html?_r=0

50. There are twenty-two of these *Double Elvis (Ferus Type)* silkscreens; one sold in 2012 at Sotheby's in New York for more than $37 million – shy of the $50 million that Sotheby's had predicted it might fetch.

51. I later learned that there was also Pino Pascali's *Machine Gun* and Gino Severinis's *Armoured Train in Action* at MOMA but saw neither.

52. Credit must go fully to Barbara Eldredge for many of these observations, outlined in her powerful postgraduate thesis: http://museummonger.files.wordpress.com/2013/01/barbaraeldredge_thesis_shortversion.pdf

53. Calvin Tompkins, *Merchants and Masterpieces: The Story of the Metropolitan Museum of Art* (New York: Dutton, 1973), p. 157.

54. http://www.metmuseum.org/collection/the-collection-online/search?when=A.D.+1900-present&deptids=4&ft=*&rpp=30&pg=2

55. Paola Antonelli, interview by Barbara Eldredge, 29 November 2012, Museum of Modern Art, New York: http://museummonger.files.wordpress.com/2013/01/barbaraeldredge_thesis_shortversion.pdf, p. 28; Paola may have changed her mind a little since, as she recently curated a MOMA exhibition called 'Design and Violence': http://designandviolence.moma.org/about/

56. Credit to Landon Y. Jones for these observations: http://www.washingtonpost.com/opinions/loaded-language-the-gun-metaphors-that-pervade-our-everyday-slang/2014/04/18/40c4053c-c3ed-11e3-b574-f8748871856a_story.html

57. According to a study conducted by researchers at Ohio State University and Annenberg Public Policy Center. The study looked at 945 movies, including the top thirty films at the box office every year from 1950 to 2012. Researchers found that, on average, violence with guns occurs more than twice an hour in

both PG-13 and R-rated pics: http://variety.com/2013/film/news/report-gun-violence-in-pg-13-movies-higher-than-r-rated-films-1200818892/

58. http://washington.cbslocal.com/2012/12/21/nra-only-way-to-stop-a-bad-guy-with-a-gun-is-with-a-good-guy-with-a-gun/

59. http://www.bjs.gov/content/pub/pdf/GUIC.PDF

60. http://www.motherjones.com/politics/2013/01/nra-hollywood-guns-movies-glock-clint-eastwood

61. Kerry Segrave, *Product Placement in Hollywood Films: A History* (Jefferson, North Carolina: McFarland & Co., 2004), p. 207.

62. http://www.motherjones.com/politics/2013/01/nra-hollywood-guns-movies-glock-clint-eastwood

63. http://pediatrics.aappublications.org/content/124/5/1495.full; they include the American Academy of Child and Adolescent Psychiatry, the American Academy of Family Physicians, the American Medical Association, the American Psychiatric Association and the American Psychological Association, in addition to the American Academy of Pediatrics.

64. http://pediatrics.aappublications.org/content/early/2013/11/06/peds.2013-1600.abstract

65. http://www.hollywoodreporter.com/news/sandy-hook-shooter-linked-violent-404576

66. https://www.youtube.com/watch?v=1j-vrxGXSXE

67. http://interpersonalresearch.weebly.com/uploads/1/0/4/0/10405979/ppmc_-_vvgs_and_real-world_violence.pdf

68. http://www.youtube.com/watch?v=GI1fNjdeOvY#t=46

Chapter 12: The Traders

1. http://www.smallarmssurvey.org/fileadmin/docs/A-Yearbook/2012/eng/Small-Arms-Survey-2012-Chapter-8-summary-EN.pdf; the trade in ammunition is estimated at some $4.26 billion.

2. The total value of the global arms trade in 2011 was estimated to be at least $43 billion. The true figure is likely to be higher: http://www.sipri.org/research/armaments/transfers/measuring/financial_values; http://www.smallarmssurvey.org/fileadmin/docs/A-Yearbook/2001/en/Small-Arms-Survey-2001-Chapter-04-EN.pdf

3. http://www.smallarmssurvey.org/fileadmin/docs/A-Yearbook/2001/en/Small-Arms-Survey-2001-Chapter-04-EN.pdf

4. http://www.boston.com/news/world/europe/articles/2005/06/30/un_report_puts_worlds_illicit_drug_trade_at_estimated_321b/

5. http://www.unodc.org/documents/human-trafficking/UNVTF_fs_HT_EN.pdf

6. According to the UN Commodity Trade Statistics Database – http://comtrade

.un.org – a rise from $434,834,395 to $1,485,328,055. Admittedly, it is almost impossible to get accurate figues on this. These figures serve as an indication only.

7. Of which about 4.5 million were firearms – 30,592,599 – as listed in: http://comtrade.un.org; http://www.smallarmssurvey.org/fileadmin/docs/A-Yearbook/2009/en/Small-Arms-Survey-2009-Chapter-01-EN.pdf/

8. Figures for 2011: http://nisatapps.prio.org/armsglobe/index.php

9. http://www.smallarmssurvey.org/fileadmin/docs/F-Working-papers/SAS-WP14-US-Firearms-Industry.pdf

10. http://shotshow.org

11. It's no surprise that the Shot Show brings in an estimated $73 million to Las Vegas.

12. http://shotshow.org/press/newsroom/

13. Of those, 51,438 are retail gun stores, 7,356 are pawn shops, 61,562 are collectors; http://abcnews.go.com/US/guns-america-statistical/story?id=17939758. The biggest dealer of all is Walmart. In 2011, in an attempt to revive flagging sales, the retailer decided to expand gun sales to just over half of its 3,982 stores nationwide – including in some urban areas. The company now sells 400 gun makes in its stores. Sales have been brisk, and the FBI received nearly 16.8 million background check requests from the store in 2012 alone. For McDonald's numbers see: http://www.theguardian.com/news/datablog/2013/jul/17/mcdonalds-restaurants-where-are-they

14. Garen J. Wintemute of the University of California carried out a survey in 2011 on the characteristics of licensed firearm retailers and retail establishments in America. Respondents to the survey had a median age of 54; 89 per cent were male; and 97.6 per cent were white. http://www.ncbi.nlm.nih.gov/pmc/articles/PMC3579296/

15. http://www.freedom-group.com/2011_10-K.pdf; specifically 'long gun sales'.

16. http://www.colt.com/Catalog/BoltActionRifles.aspx

17. http://www.redfield.com/redfield-revenge/

18. For a fascinating account of the rise of Glock, read Paul Barrett on the same: http://www.amazon.com/Glock-The-Rise-Americas-Gun/dp/0307719952

19. Sig Sauer, Beretta and Taurus: http://www.smallarmssurvey.org/fileadmin/docs/F-Working-papers/SAS-WP14-US-Firearms-Industry.pdf

20. http://www.bbc.co.uk/worldservice/learningenglish/language/wordsinthenews/2014/01/140113_witn_kalashnikov.shtml

21. http://www.cityam.com/207632/russias-kalashnikov-be-made-us-after-sanctions-import-ban

22. http://www.jcf.gov.jm/article/statement-police-high-command-0; http://bigstory.ap.org/article/jamaica-police-charged-killing-pregnant-woman

23. http://www.jamaicaobserver.com/NEWS/DPP-rules-that-accused-St-Thomas-killer-cop-be-charged-with-murder; http://library.jcsc.edu.jm/xmlui/bitstream/

handle/1/31/Forcepercent20Orderspercent203281Apercent202010-04-22.pdf
?sequence=1

24. http://us.glock.com/products/sector/law-enforcement

25. Data from http://www.fatalencounters.org

26. Richard Jones, *Jane's Infantry Weapons 2009–2010* (London: Jane's Information Group, 2009), p. 897; http://www.hrw.org/world-report/2014/country-chapters/iraq; http://www.globalsecurity.org/intell/world/iraq/ips.htm

27. https://www.youtube.com/watch?v=1VEykPjHWp8&feature=related; http://www.uasvision.com/2012/09/12/belarus-bypasses-eu-sanctions-to-buy-german-engines-for-uas/; https://freedomhouse.org/fair-play-beyond-sports/timeline-20-years-human-rights-abuses-belarus; https://euobserver.com/foreign/117489

28. http://en.apa.az/news/204647; http://www.hrw.org/europecentral-asia/azerbaijan

29. http://www.jpost.com/Israel/We-had-no-choice; http://www.isayeret.com/guides/weapons.shtml; http://www.hrw.org/middle-eastn-africa/israel-palestine

30. http://indecom.gov.jm/2013%20Statistics%20Press%20Release.pdf

31. http://www.amnesty.org/en/region/jamaica/report-2011

32. The Glock USA sale to Jamaica was legal and passed export controls.

33. http://newsinfo.inquirer.net/267292/pnp-nears-p1-b-deal-to-buy-60000-guns

34. http://www.thefirearmblog.com/blog/2013/06/27/philippine-national-police-buy-14000-glock-17-pistols/

35. http://www.nytimes.com/2014/12/04/world/asia/report-condemns-police-torture-in-philippines.html?_r=0; http://www.hrw.org/news/2014/06/16/statement-extrajudicial-killings-philippines

36. http://www.thenational.ae/news/world/asia-pacific/philippine-police-warned-dont-sell-off-your-brand-new-glock; http://www.cordilleraism.com/2014/08/policeman-kills-two-men-in-shooting.html

37. It had the serial number PNP 03283: http://www.baguiomidlandcourier.com.ph/front.asp?mode=archives/2014/august/8-10-2014/front3.txt

38. http://www.bbc.co.uk/news/world-latin-america-28291070

39. http://international.sueddeutsche.de/post/93766223555/the-dubious-colombia-exports-of-german-arms; http://www.icij.org/blog/2014/08/tracking-cross-border-weapons-trade-youtube

40. http://international.sueddeutsche.de/post/93766223555/the-dubious-colombia-exports-of-german-arms

41. http://www.theguardian.com/world/2014/feb/25/oberndorf-german-town-armed-world-heckler-kock

42. http://www.hrw.org/news/2011/11/09/mexico-widespread-rights-abuses-war-drugs

43. http://www.theguardian.com/world/2014/feb/25/oberndorf-german-town-armed-world-heckler-kock

44. Freedom of Information requests are often denied when related to matters of national security or business.

45. http://www.sipri.org/research/disarmament/eu-consortium/publications/publications/non-proliferation-paper-7

46. http://test.pmddtc.state.gov/reports/congnotices/113/12-164.pdf

47. http://www.smallarmssurvey.org/fileadmin/docs/A-Yearbook/2009/en/Small-Arms-Survey-2009-Chapter-01-EN.pdf

48. http://books.sipri.org/product_info?c_product_id=476#

49. In China there are four state-owned enterprises authorised to export guns – Norinco, Poly Technologies, Jing'an Import and Export Corporation, and China Xinxing Import and Export Company.

50. Based on UN Comtrade and estimates.

51. Several other states in Asia have imported guns from China, with Pakistan and Bangladesh being the most prominent, and there has been a reported rise in sales to Latin America in recent years. In the Middle East, Egypt, Jordan, Lebanon and Qatar also imported guns from China between 2006 and 2010. Iran has also been a major recipient of Chinese guns since the Iran–Iraq War, though there have been recent reports suggesting they have since wound down their arms sales there.

52. http://books.sipri.org/files/PP/SIPRIPP38.pdf

53. http://www.voanews.com/content/china-says-it-was-unaware-arms-dealers-met-with-gadhafi-representatives-129252233/144797.html

54. Chinese officials acknowledged that the meeting took place and a delegation from the Gaddafi regime visited Chinese arms companies in Beijing in July 2011 to discuss arms contracts.

55. http://www.poa-iss.org/CASAUpload/ELibrary/S-2008-371-Ex-Liberia-E.pdf

56. Rosoboronexport announced a record number of contracts in 2012: http://vz.ru/news/2013/2/13/620142.html

57. http://www.smallarmssurvey.org/fileadmin/docs/A-Yearbook/2013/en/Small-Arms-Survey-2013-Chapter-08-Annexes-8.1-8.2-EN.pdf

58. http://english.pravda.ru/news/russia/01-03-2001/39602-0/

59. http://cms.privatelabel.co.za/pls/cms/iac.page?p_t1=2245&p_t2=0&p_t3=0&p_t4=0&p_dynamic=YP&p_content_id=227743&p_site_id=113

60. In 1993, US President Bill Clinton said: 'The situation in Angola constitutes a threat to international peace and security'; arms transfers to Angola were also subject to a UN Security Council Resolution 864, barring certain transfers.

61. Data available on www.equasis.com; https://www.unodc.org/documents/data-and-analysis/tocta/6.Firearms.pdf

62. http://www.bbc.co.uk/news/world-africa-14095300

63. http://www.slideshare.net/RobSentseBc/the-odessa-network; http://www.bbc.co.uk/news/world-africa-13286306

64. http://nisatapps.prio.org/armsglobe/index.php

65. http://www.nytimes.com/2005/07/16/international/europe/16ammo.html?pagewanted=all&_r=0

66. C. J. Chivers, *The Gun* (Harmondsworth: Penguin 2010), p. 368.

67. Rachel Stohl and Suzette Grillot, *The International Arms Trade* (Cambridge: Polity Press, 2012).

68. http://molodyvcheny.in.ua/files/journal/2014/8/26.pdf

69. http://www.kyivpost.com/content/ukraine/ukraine-worlds-4th-largest-arms-exporter-in-2012-according-to-sipri-321878.html

70. http://polemika.com.ua/news-29257.html; http://www.globalinitiative.net/wpfb-file/c4ads-the-odessa-network-mapping-facilitators-of-russian-and-ukranian-arms-transfers-sept-2013-pdf/

71. http://www.oktport.com.ua/index.php?view=image&format=raw&type=img&id=13&option=com_joomgallery&Itemid=5

72. http://www.washingtonpost.com/world/national-security/ukrainian-port-eyed-as-analysts-seek-syrias-arms-source/2013/09/07/f61b0082-1710-11e3-a2ec-b47e45e6f8ef_story.html

73. http://www.businessinsider.com/putin-ukraine-and-syria-2014-3

74. http://www.washingtontimes.com/news/2014/apr/7/hannaford-the-case-of-the-misreported-ship/#ixzz3Di677xCK

75. Their fleet includes LS *Aizenshtat*, *Ocean Fortune*, *Ocean Winner*, *Ocean Force*, *Ocean Energy* and *Ocean Power*.

76. http://ylpr.yale.edu/inter_alia/collection-gap-explained-white-collar-practitioners-view

77. http://s3.documentcloud.org/documents/1164991/originalc4adscomplaint.txt

Chapter 13: The Smugglers

1. http://www.liveleak.com/view?i=cc2_1384056914; http://www.securitycouncilreport.org/atf/cf/%7B65BFCF9B-6D27-4E9C-8CD3-CF6E4FF96FF9%7D/s_2014_106.pdf; intriguingly, despite official statements stating that there were arms on board, the Security Council Report concluded that 'no arms were onboard', despite reports and photographs such as this: http://www.news.com.au/world/cargo-ship-nour-m-stopped-and-20000- kalashnikov-assault-rifles-found/story-fndir2ev-1226756397670 and here http://www.jacarandafm.com/post/greek-coastguard-halts-ship-carrying-20000-kalashnikovs/

2. http://www.keeptalkinggreece.com/2013/11/09/greek-coastguard-intercepts-sierra-leone-flagged-vessel-with-20000-kalashnikov-rifles

3. The Greek Coastguard issued a statement saying: 'The exact destination of the arms and ammunition has yet to be verified.' The captain and seven of his crew, two other Turks and five Indians, were arrested. It should be noted, though, that the information systems from which the Greek media get their data do also warn that their data is not official and cannot be used for commercial or navigation purposes. It would, therefore, be presumptuous to come to hard conclusions without an official report.

4. http://www.vesselfinder.com/news/1700-VIDEO-Cargo-ship-Nour-M-sinks-in-Rhodes-due-to-bad-weather

5. Intriguingly, perhaps, there was some suggestion on chat boards that the Greek authorities considered giving the seized rifles to the Greek National Guard (see here: http://www.militaryphotos.net/forums/showthread.php?232274-Greece-stops-ship-carrying-20-000-Kalashnikov-guns). If this is the case, then the Security Council were misled when they were told that there were no rifles (see above). Perhaps, given Greece's financial crisis, the government saw this as an easy way to arm themselves. Or perhaps there were no weapons to begin with, and the authorities, the media or both misspoke.

6. http://www.lefigaro.fr/international/2010/10/26/01003-20101026AR TFIG00809-odessa-fait-de-la-resistance.php

7. http://www.nytimes.com/2003/08/17/magazine/arms-and-the-man.html?page wanted=all&src=pm

8. http://www.gunpolicy.org/firearms/region/cp/ukraine

9. Andrew Feinstein, *The Shadow World: Inside the Global Arms Trade* (London: Penguin, 2012), p. 118.

10. So renowned did Ukraine's merchants of death become that Nicolas Cage's character in the film *Lord of War* was made into a Ukrainian-American arms dealer, even though the character was based on the arms dealer Victor Bout, who was Russian.

11. http://www.guardian.co.uk/world/2011/nov/02/viktor-bout-arms-trade; http://www.guardian.co.uk/world/2001/jul/09/armstrade.iantraynor

12. http://www.pravda.com.ua/rus/news/2003/02/3/4371494/; http://www.pbs.org/frontlineworld/stories/sierraleone/minin.html

13. In 1992, when arms dealer Monzer al-Kassar wanted to ship arms into Croatia, he presented a Polish supplier with a certificate that was supposed to have been issued by the People's Democratic Republic of Yemen, even though that country had not existed for two years: http://www.newyorker.com/maga zine/2010/02/08/the-trafficker

14. http://www.un.org/News/Press/docs/2006/dc3029.doc.htm

15. http://www.economist.com/news/europe/21604234-fight-against-corruption-steep-uphill-battle-ostrich-zoo-and-vintage-cars

16. Aaron Karp, 'Completing the Count: Civilian Firearms – Annexe Online', *Small Arms Survey 2007: Guns and the City* (Cambridge: Cambridge University

Press, 2007), Chapter 2 (Annexe 4), p. 67. Between 2004 and 2011, the United Nations Monitoring Group reported almost 50,000 instances involving the transfer of small arms and light weapons in Somalia.

17. http://www.smallarmssurvey.org/fileadmin/docs/A-Yearbook/2007/en/full/Small -Arms-Survey-2007-Chapter-08-EN.pdf

18. http://www.smallarmssurvey.org/fileadmin/docs/A-Yearbook/2013/en/Small -Arms-Survey-2013-Chapter-11-EN.pdf

19. Ibid.

20. Ibid.

21. C. J. Chivers, *The Gun* (Harmondsworth: Penguin, 2010), p. 381.

22. Though, reportedly, thirteen people still had to have amputations following the punishment: L. E. Graham and R. C. Parke, 'The Northern Ireland Troubles and Limb Loss: A Retrospective Study', *Prosthetics and Orthotics International* 28, 3, 2004, pp. 225–9.

23. Loose Minute to PS. US of S Northern Ireland, Losses of UDR Weapons – Comparison of Losses of Weapons Between Battalions of UDR and Incidents of Suspected Collusion, 1 August 1972.

24. http://www.theguardian.com/uk/2012/oct/15/uk-arms-northern-ireland-loyalist-massacre. The *Guardian* reported that 'there are serious concerns about the way the Loughinisland killings were investigated, with a subsequent inquiry by the police ombudsman establishing that police failed to take some suspects' fingerprints or DNA samples. Police have admitted that one key piece of evidence – the getaway car – was destroyed. There is no evidence that any officer sought or gave permission for this to be done.'

25. Ibid.

26. http://www.independent.co.uk/news/uk/catholic-shot-dead-in-ambush-attack-on-workmen-in-minibus-seen-as-loyalist-warning-to-workers-at-shorts-aerospace-factory-1510516.html. http://republican-news.org/current/news/2011/07/new _revelation_points_to_lough.html

27. http://www.belfastdaily.co.uk/2012/10/16/loughinisland-pub-massacre-secret-gun-shipment-funded-by-british-army/

28. http://www.theguardian.com/uk/2012/oct/15/northern-ireland-loyalist-shootings-loughinisland

29. Sean Boyne, *Gunrunners: The Covert Arms Trail to Ireland* (Dublin: O'Brien, 2006), p. 368.

30. Thomas McErlean, John Murray and IRA volunteer Caoimhín Mac Brádaigh.

31. http://republican-news.org/archive/2003/March20/20mont.html

32. http://www.telegraph.co.uk/news/worldnews/northamerica/usa/10249892/Jailed-Boston-mobsters-gang-smuggled-weapons-to-the-IRA-in-coffins.html

33. http://www.nytimes.com/2013/02/26/world/middleeast/in-shift-saudis-are -said-to-arm-rebels-in-syria.html?partner=rss&emc=rss&smid=tw-nytimes &_r=1&

34. The allegations were made by José 'Pepe' Grinda Gonzalez, Spain's national court prosecutor. Gonzalez was responsible for the investigation into Zakhar Kalashov, reportedly the most senior mafia figure to be jailed outside Russia. http://www.theguardian.com/world/us-embassy-cables-documents/247712?guni=Article:inper cent20bodyper cent20link

35. http://www.reuters.com/article/2014/01/27/us-usa-syria-rebels-idUS-BREA0Q1S320140127

36. http://www.npr.org/templates/story/story.php?storyId=1262079

37. http://www.telegraph.co.uk/news/worldnews/asia/afghanistan/11004928/Afghanistan-has-cost-more-to-rebuild-than-Europe-after-Second-World-War.html

38. http://www.globalpost.com/dispatch/news/regions/americas/united-states/140509/the-us-military-describes-its-mistakes-afghanistan

39. http://rt.com/news/176004-afghanistan-missing-weapons-sigar/

40. http://www.sigar.mil/pdf/Audits/SIGAR-14-84-AR.pdf

41. http://www.telegraph.co.uk/news/worldnews/asia/afghanistan/5355937/Taliban-using-ammunition-from-Afghan-army.html

42. http://www.washingtonpost.com/wp-dyn/content/article/2007/08/05/AR2007080501299.html; http://www.gao.gov/new.items/d07711.pdf. The Department of Defense and the MNF-I (Multinational Force-Iraq) could not fully account for Iraq security forces' receipt of US-provided equipment. This was put down to not having a centralised record of all equipment, insufficient number of staff, not collecting documents that confirmed when the equipment was received and the sheer quantities of equipment delivered.

43. http://www.smallarmssurvey.org/fileadmin/docs/A-Yearbook/2007/en/full/Small-Arms-Survey-2007-Chapter-03-EN.pdf

44. http://www.dailymail.co.uk/news/article-2749197/ISIS-arming-US-military-hardware-wage-jihad-Middle-East-seizing-weapons-Syrian-rebels-Iraqi-soldiers.html

45. http://conflictarm.com/images/dispatch_iraq_syria.pdf

46. http://justiceinmexico.files.wordpress.com/2013/02/130206-dvm-2013-final.pdf

47. http://www.nbcnews.com/id/46425305/ns/world_news-americas/t/mexico-president-felipe-calderons-message-us-no-more-weapons/#.VB15cUvobx4

48. http://www.bbc.co.uk/news/world-latin-america-20825061

49. In 2010–12 the various estimated figures of smuggled guns ranged between 106,700 and 426,729. This constitutes a 187 per cent increase from the numbers trafficked between 1997 and 1999: http://catcher.sandiego.edu/items/peacestudies/way_of_the_gun.pdf; see also http://www.bbc.co.uk/news/world-latin-america-20825061

50. http://ncronline.org/news/global/us-gun-policy-no-longer-domestic-weapons-are-smuggled-mexico

51. Data from the University of San Diego Trans-Border Institute. Its figures

range between 39.4 and 52.7 per cent: http://catcher.sandiego.edu/items/peacestudies/way_of_the_gun.pdf

52. Data from the University of San Diego Trans-Border Institute. The US Justice Department's Bureau of Alcohol, Tobacco, Firearms and Explosives reported in 2012 that of the more than 99,000 guns that Mexican authorities seized and submitted to the ATF for tracing between 2007 and 2011, 68,000 came from the US. There has, though, been some debate over these figures. On the one hand, the US Government Accountability Office estimated 90 per cent or more of guns used in Mexico were from the US. On the other hand some say only 17 per cent of weapons found at Mexican crime scenes originate from the US – that, of 29,000 firearms recovered at crime scenes in Mexico, 23,886 could not be traced to the US. This second argument is misleading. Many of the 29,000 firearms were not even sent for tracing, and so we have no idea if they were, or were not, from the US. Also, obliterated serial numbers, incomplete sales record-keeping and private purchases make it hard to find the origin of a gun. To this end, the ATF figures seem logical and acceptable. See also: http://www.bbc.co.uk/news/world-latin-america-20825061

53. http://www.csmonitor.com/World/Americas/Latin-America-Monitor/2011/0708/Mexican-cartel-leader-claims-gang-buys-all-its-guns-in-US

54. http://ncronline.org/news/global/us-gun-policy-no-longer-domestic-weapons-are-smuggled-mexico

55. http://www.un.org/events/smallarms2006/pdf/factsheet_1.pdf

56. http://iis-db.stanford.edu/evnts/6716/Oeindrila_Dube,_Cross_Border_Spillover.pdf

57. In the four years following the lapse of America's assault weapons ban in 2004, 60,000 illegal firearms seized in Mexico were traced back to the US.

58. https://www.atf.gov/sites/default/files/assets/statistics/TraceData/TraceData_Intl/2013/central_america_-_cy_2013.pdf

59. Ian Thomson, *The Dead Yard* (London: Faber and Faber, 2009), p. 316.

60. What has muddied the waters about the impact of straw purchasing on the Mexican and Central American markets was the botched sting operation 'Fast and Furious' by the ATF. There the US authorities lost track of 1,400 'planted' firearms that they were hoping would lead them to drug leaders, but the guns ended up arming Mexican gangs such as the Sinaloa cartel.

61. http://www.vpc.org/florida/FLNeedlemanComplaint080410.pdf

62. http://articles.sun-sentinel.com/2008-08-08/news/0808070489_1_gun-shop-gun-range-felons

63. Investigation by the pro-gun control group Mayors Against Illegal Guns.

64. Gun rights activists say that figure is grossly overstated and based on outdated data.

65. http://www.atf.gov/sites/default/files/assets/Firearms/chap1.pdf. A separate pro-gun study concluded gun shows had little effect on deaths in the weeks

following the shows. It found tighter California gun show regulations did not, for instance, reduce firearm-related deaths there. But studies have shown that gun crimes rarely involve weapons bought shortly before a crime's occurrence. And looking at crimes solely within 25 miles of a gun show ignores findings about the geography of illegal gun markets; roughly two-thirds of gun crimes are from firearms purchased out of state or far away from the scene of the deed.

66. http://usatoday30.usatoday.com/news/opinion/2001-12-13-nceditf.htm

67. http://www.nytimes.com/2001/11/13/us/nation-challenged-gun-control-gun-foes-use-terror-issue-push-for-stricter-laws.html

68. http://www.washingtonpost.com/opinions/closing-the-terror-gap-and-the-gun-show-loophole/2011/06/06/AGTKubKH_story.html. Perhaps the most notable non-international case, though, was the Columbine High School massacre. The two perpetrators bought two shotguns and a Hi-Point semi-automatic from a private seller for cash at the Tanner Gun Show in Adams County, Colorado. No questions were asked, and no paperwork was filled out. A fourth gun was also purchased directly by the boys from a private seller, who in turn had got it from an unlicensed seller at the Tanner Gun Show.

69. http://www.buzzfeed.com/juangastelum/militia-leader-calls-on-members-to-go-armed-to-the-border#6d6htr

70. http://www.theguardian.com/world/2014/jul/09/central-america-child-migrants-us-border-crisis

71. http://www.newrepublic.com/article/118759/nra-and-gun-trafficking-are-adding-fuel-border-migrant-crisis

Chapter 14: The Lobbyists

1. As Human Rights Watch was to state: 'It was regularly resupplied by air from South Africa, to the extent that it was better equipped than the Mozambique army.' See: http://www.hrw.org/sites/default/files/reports/Mozamb927.pdf

2. http://www.bbc.co.uk/news/world-africa-13890416

3. http://vimeo.com/51739769

4. http://pedroreyes.net/palasporpistolas.php

5. http://www.jewelryforacause.net/our-collections/caliber-collection; http://www.thedailybeast.com/articles/2013/01/25/cory-booker-s-murder-accessory-turning-buyback-guns-into-jewelry.html

6. http://www.fonderie47.com/products/inversion-principle

7. http://allafrica.com/view/group/main/main/id/00030721.html

8. http://www.the-monitor.org/index.php/publications/display?url=lm/2013/sub/Major_Findings.html

9. It was the largest amount ever given. Of this, $32 million was for victim assistance and $15 million spent on advocacy in 2012.

10. The treaty covered all forms of major weapons – battle tanks, artillery, combat aircraft, warships and missiles – as well as guns.

11. http://www.reuters.com/article/2013/03/29/us-arms-treaty-un-idUS-BRE92R10E20130329

12. Of the fifteen top arms gun exporters, the following have signed but not yet ratified the Arms Trade Treaty: Brazil, Turkey and the United States. India, China, North Korea, Pakistan, the Russian Federation and Canada have not even signed up to it. So Austria, Belgium, Germany, Italy, Switzerland and the United Kingdom are the only major gun-producing countries who are signatories and have ratified it; http://disarmament.un.org/treaties/t/att; http://www.un.org/disarmament/ATT/

13. Its full name is the Programme of Action to Prevent, Combat and Eradicate the Illicit Trade in Small Arms and Light Weapons in All Its Aspects.

14. All of the other eleven top producers: Brazil, Canada, Germany, India, Italy, Pakistan, the Russian Federation, Switzerland, Turkey, the United Kingdom and the United States did file a report; http://www.poa-iss.org/Poa/National ReportList.aspx

15. http://www.reachingcriticalwill.org/images/documents/Disarmament-fora/salw/poa-assessment.pdf

16. This tries to do concrete things about stopping the illicit spread of guns. Things like stamping out illegal manufacture; making sure that governments have databases about gun serial numbers; making sure that gun companies have proper security to stop people stealing arms off the manufacturing line; making sure that destroyed guns really are destroyed; and sharing information with other international agencies about gun smugglers.

17. Those who have signed it and have yet to ratify it include Austria, Canada, China, Germany, India and the United Kingdom. Those who never even signed it include North Korea, Pakistan, the Russian Federation, Switzerland and the United States: http://www.unodc.org/unodc/en/treaties/CTOC/countrylist-firearmsprotocol.html; http://www.smallarmssurvey.org/weapons-and-markets/producers/industrial-production.html

18. There is also the Wassenaar Arrangement. Signed by forty-one countries, it seeks to make the transfer of guns and other weapons more transparent. It does this by trying to establish a baseline as to when a state should decline a small-arms export licence. This was an attempt to stop things like guns fuelling terrorism, ongoing conflict and human rights abuses. It has been reasonably effective in increasing transparency around arms transfers, but there have been endless debates and arguments. How much information should be shared? How do you classify a state as being 'of concern'? Is this shipment of guns really going to be a 'desta-

bilising transfer'? And, of course, there is the fact that Belarus, China and Israel – significant gun producers – haven't signed up at all. These hard realities are repeated with the UN Conventional Arms Register. This seeks to prevent large-scale build-ups of arms. It's mainly about big weapons like tanks and missile launchers, but there is a section on guns, inviting states to provide information on things like gun exports, destinations, stockpiles and manufactured numbers. But its voluntary nature means that in 2012 only seventy-two countries reported.

19. http://www.hsph.harvard.edu/hicrc/firearms-research/guns-and-death

20. David B. Kopel, 'Gun Ownership and Human Rights', *Dialogue* 9, 2, Winter/ Spring 2003, p. 7, http://www.davekopel.com/2A/Foreign/Brown-Journal-Kopel.pdf; whether a well equipped and armed state – with airpower and tanks – could be prevented from human rights violations by small arms alone is debatable.

21. http://www.cfi-icaf.ca/index.php?option=com_content&view=article&id=578 :canada-tries-to-exempt-hunting-rifles-from-un-arms-trade-treaty-negotiations -&catid=79:the-toronto-sun&Itemid=113; though when I contacted Survival International, a charity dedicated to the rights of indigenous peoples, they said: 'It really hasn't come up as an important issue' (personal correspondence with Stephen Corry at Survival International).

22. http://www.wfsa.net/about.html

23. http://www.nraontherecord.org/john-bolton/

24. http://www.smallarmssurvey.org/fileadmin/docs/L-External-publications/2002 /2002-Krause-MultilateralDiplomacyNormBuilding.pdf

25. http://gunowners.org/a07152011.htm

26. http://www.amnestyusa.org/news/press-releases/nra-must-drop-its-campaign-of -lies-against-un-global-arms-trade-treaty

27. http://www.foxnews.com/world/2012/07/20/un-arms-treaty-aims-at-terror-but-puts-second-amendment-in-crosshairs/#ixzz21dn2xSrx

28. http://www.snopes.com/politics/guns/untreaty.asp

29. http://www.ammoland.com/2013/08/obama-will-sign-un-gun-treaty-while-congress-is-on-vacation/

30. http://www.americanbar.org/content/dam/aba/administrative/individual_ rights/aba_chr_white_paper_att_final.authcheckdam.pdf

31. http://www.reuters.com/article/2013/09/25/us-un-assembly-kerry-treaty-idUSBRE98O0WV20130925

32. http://www.washingtonpost.com/world/national-security/us-nra-square-off-over-small-arms-treaty/2013/03/16/ae495dae-8d76-11e2-b63f-f53fb9f2fcb4_ story.html

33. http://thehill.com/blogs/floor-action/senate/290001-senate-votes-to-stop-us-from-joining-un-arms-treaty

34. http://fas.org/sgp/crs/weapons/R42678.pdf

35. http://www.cbc.ca/news/canada/nra-involved-in-gun-registry-debate-1.923766

36.   http://www.americasquarterly.org/content/nras-hemispheric-reach

37.   http://www.insightcrime.org/news-analysis/brazil-police-say-sea-is-new-arms-trafficking-frontier

38.   http://www.nytimes.com/2006/09/17/magazine/17wwln_essay.html

39.   http://news.bbc.co.uk/1/hi/world/americas/4368598.stm?words=Breivik

40.   In one speech he spoke to the hundreds of thousands of 'good Americans' who, he claimed, will see his speech online. Six months on only 7,517 had watched it on the NRA official YouTube channel: https://www.youtube.com/watch?v=BCFB0N_jzMk

41.   http://www.gallup.com/poll/159578/nra-favorable-image.aspx

42.   http://www.theguardian.com/commentisfree/2013/sep/21/american-gun-out-control-porter

43.   http://www.youtube.com/watch?v=_jm2IBUZxZ0&index=7&list=PLyaSPxNidLLvIef8u5rR-6siU2Bs8csn0

44.   http://dailycaller.com/2014/03/06/cpac-wayne-lapierres-speech/2/

45.   http://www.nraila.org/news-issues/fact-sheets/2013/more-guns-less-crime-2013.aspx

46.   http://www.theguardian.com/commentisfree/2014/apr/28/nra-war-on-america-wayne-lapierre-indianapolis

47.   http://www.usatoday.com/story/news/politics/2013/05/04/nra-meeting-la pierre-membership/2135063/; http://www.businessinsider.com/gun-industry-funds-nra-2013-1?IR=T

48.   http://www.vpc.org/studies/bloodmoney2.pdf

49.   http://www.campaign2unload.org/wp-content/uploads/2014/04/Gun-Industry-Members.pdf

50.   It led Sturm, Ruger and Co. to pledge $1 for each new Ruger firearm sold between the 2011 and 2012 NRA Annual Meetings. Their goal was a cheque for $1 million to the NRA. Their CEO got his golden jacket – they made over $1.2 million.

51.   In addition the National Association of Gun Rights spent nearly $6.8 million and the National Shooting Sports Foundation, over $2.3 million; http://www.opensecrets.org/lobby/clientsum.php?id=D000000082&year=2013

52.   Dennis A. Henigan, *Lethal Logic: Exploding the Myths that Paralyze American Gun Policy* (Dulles, VA: Potomac Books, 2009), p. 184.

53.   http://www.businessinsider.com/smith-and-wesson-almost-went-out-of-business-trying-to-do-the-right-thing-2013-1

54.   Ibid.

55.   http://www.telegraph.co.uk/news/politics/ukip/10595087/Hand-guns-should-be-legalised-and-licensed-Nigel-Farage-has-said.html; http://www.politics.co.uk/news/2014/01/27/police-shoot-down-nigel-farage-s-handgun-call

56.   http://www.washingtonpost.com/world/europe/germany-initiates-new-gun-registry/2013/01/19/86bb29f2-60da-11e2-b05a-605528f6b712_story.html

57. http://www.nraila.org/news-issues/articles/2013/10/secretary-general-of-interpol-suggests-an-armed-citizenry-to-combat-mass-violence.aspx

58. http://www.state.gov/j/ct/rls/crt/2013/224833.htm

59. http://www.cdc.gov/nchs/fastats/injury.htm

60. http://www.propublica.org/article/democrats-push-to-restart-cdc-funding-for-gun-violence-research

61. http://www.nejm.org/doi/full/10.1056/NEJM198606123142406

62. http://www.npr.org/blogs/itsallpolitics/2013/01/14/169164414/lack-of-up-to-date-research-complicates-gun-debate

63. http://www.nytimes.com/2011/01/26/us/26guns.html

64. Ibid.

65. In the aftermath of the Sandy Hook massacre, President Obama issued a Presidential Memorandum directing the CDC and others to conduct research into the causes and prevention of crime. He called upon Congress to allocate $10 million for the CDC to conduct further research, including the relationship between video games, media images and violence. Despite media reports that the funding freeze would be lifted in 2014, at the time of writing it was not clear that this will happen any time soon: http://www.msnbc.com/rachel-maddow-show/cdc-still-cant-get-funding-research

66. The lobbyists also turned their attention to the US National Institutes of Health (NIH) – a body that invests $30 billion into medical research every year. A 2011 act had a rider that stated no NIH funding 'may be used, in whole or part, to advocate or promote gun control'.

67. Ulysses S. Grant, Theodore Roosevelt, William Howard Taft, Dwight D. Eisenhower, John F. Kennedy (the lone Democrat), Richard M. Nixon, Ronald Reagan and George H. W. Bush. The elder Bush resigned from the NRA in 1995 after LaPierre's attack on federal agents in the wake of the Oklahoma City tragedy: http://www.nytimes.com/1995/05/11/us/letter-of-resignation-sent-by-bush-to-rifle-association.html

68. http://www.theguardian.com/world/2013/apr/18/pro-gun-groups-donated-senators

69. http://www.publicintegrity.org/2013/05/01/12591/gun-lobbys-money-and-power-still-holds-sway-over-congress

70. http://www.cpsc.gov/en

71. http://www.usatoday.com/story/news/nation/2014/01/27/guns-children-hospitalizations/4796999/

72. http://www.usatoday.com/story/news/nation/2013/04/11/guns-child-deaths-more-than-cancer/2073259/

73. They were later to change the date: http://www.huffingtonpost.co.uk/2013/12/16/guns-save-lives-day_n_4452075.html

74. http://news.bbc.co.uk/1/hi/uk/7056245.stm

75. For more on this, visit: https://aoav.org.uk/2014/15-years-since-columbine/

76. http://www.washingtonpost.com/opinions/when-australians-gave-back-their-guns/2013/08/23/108458dc-0c09-11e3-8974-f97ab3b3c677_story.html

77. It is not just mass shootings that have been impacted. At that time Australia's firearm mortality rate per population was 2.6 per 100,000. Today the rate is under 1 in 100,000, less than one-tenth the US rate. Sources: the Australian Bureau of Statistics and the US Centers for Disease Control and Prevention. These are for all gun deaths – homicide, suicide and unintentional. If you just focus on gun homicide rates, the US outstrips Australia thirty-fold.

78. http://www.pbs.org/wgbh/pages/frontline/social-issues/newtown-divided/how-the-gun-rights-lobby-won-after-newtown/

79. http://www.huffingtonpost.com/2013/10/29/guns-in-school_n_4174071.html

80. http://www.abc.net.au/news/2014-06-08/us-rush-on-guns-triggers-shooting-supplies-shortage/5507752

81. http://www.dailymail.co.uk/news/article-2251762/NRA-condemned-astonishing-response-Sandy-Hook-massacre-calling-schools-arm-themselves.html

82. http://www.bbc.co.uk/news/world-us-canada-20815130

83. http://www.businessinsider.com/nra-practice-range-app-2013-1

84. After much criticism, they raised the age to twelve years; http://bigstory.ap.org/article/nra-shooting-game-no-longer-preschoolers

85. http://www.latimes.com/nation/nationnow/la-na-nn-gun-violence-schools-20140610-story.html; http://www.businessinsider.com/fbi-says-mass-shootings-are-on-the-rise-in-america-2014-9?IR=T; Everytown.org, a pro-gun law lobby group, listed 72 incidents since Sandy Hook, so the *LA Times* figure is a conservative one.

86. http://www.telegraph.co.uk/news/worldnews/northamerica/usa/10889499/Bulletproof-blankets-to-protect-children-from-tornadoes-and-crazed-gunmen.html

87. Watts spent a number of weeks in 1945 in Obersalzberg or Eagle's Nest, Hitler's Berghof. While there, he hunted not for Nazis, but for treasure. He collected thousands of items of silverware, uniforms and documents.

88. http://www.youtube.com/watch?v=bQWb-5nblx4

89. http://www.zombiesurvivalcourse.com

90. http://www.outdoorlife.com/photos/gallery/survival/2010/03/surviving-undead-zombie-guns

91. http://www.cinemablend.com/television/Walking-Dead-Season-4-Premiere-Crushes-With-16-1-Million-Viewers-59862.html

92. *Dead Snow* is a 2009 film about a ski vacation that goes a bit wrong for a group of medical students as they find themselves confronted by an 'unimaginable menace'. It cost about $800,000 to make and made about $2 million at the box office alone.

93. http://www.huffingtonpost.com/2013/03/11/zombie-fads_n_2852032.html

94. http://www.abc-7.com/story/25409465/abc7-extra-zombies-and-guns-in-fla

95. http://www.palgrave-journals.com/ipr/journal/v1/n2/abs/ipr201313a.html

96. http://www.foreignpolicy.com/articles/2014/05/13/exclusive_the_pentagon_has
     _a_plan_to_stop_the_zombie_apocalypse

97. http://www.theguardian.com/commentisfree/2013/sep/21/american-gun-out-
     control-porter

98. http://www.youtube.com/watch?v=a2gCFOtaZPo

99. http://www.vpc.org/studies/militarization.pdf

100. http://www.nrapublications.org/index.php/11920/obamas-secret-plan-to-
     destroy-the-second-amendment-by-2016/

101. http://www.prweb.com/releases/2013/11/prweb11296403.htm

102. To 2.78 million in December, an all-time record and a 49 per cent increase
     over December 2011.

103. G. J. Wintemute, 'Guns, Fear, the Constitution, and the Public's Health',
     N. Engl. J. Med. 358, 2008, pp. 1421–4. Such conclusions are consistent with
     a study from the Violence Policy Center, based on 2008 data from the CDC.
     It found: 'States with higher gun ownership rates and weak gun laws have
     the highest rates of gun death . . . The analysis reveals that the five states
     with the highest per capita gun death rates were Alaska, Mississippi, Louisiana,
     Alabama and Wyoming. Each of these states had a per capita gun death rate
     far exceeding the national per capita gun death rate of 10.38 per 100,000 for
     2008. Each state has lax gun laws and higher gun ownership rates. By contrast,
     states with strong gun laws and low rates of gun ownership had far lower
     rates of firearm-related death.'

Chapter 15: The Manufacturers

1. http://www.amazon.co.uk/Moby-Dick-Herman-Melville/dp/1494316641

2. Reuters was set up the month before.

3. Samuel Colt, 'On the Application of Machinery to the Manufacture of
    Rotating Chambered-Breech Fire-Arms, and the Peculiarities of those Arms',
    Minutes of Proceedings of the Institution of Civil Engineers, vol. XI, session
    1851–2 (London: Institution of Civil Engineers, 1852), http://books.google
    .com/ebooks/reader?id=QnkDAAAAYAAJ&printsec=frontcover&output=read
    er&pg=GBS.PA13. Again, thanks to Barbara Eldredge for alerting me to this
    meeting in her excellent thesis: http://gundesigndotorg.files.wordpress
    .com/2013/05/barbaraeldredge_missingthemoderngun.pdf

4. This was not a 'new' idea – the design had been around for a few years. What
    was significant was its presentation as a manufacturing design upon the world
    stage.

5. As Colt said on that day: 'When a new piece is required, a duplicate can be supplied with greater accuracy and less expense, than could be done by the most skilful manual labour, or on active service a number of complete arms may be readily made up from portions of broken ones, picked up after an action.'

6. http://gundesigndotorg.files.wordpress.com/2013/05/barbaraeldredge_missing themoderngun.pdf

7. Mass production using interchangeable parts was actually first achieved in 1803 by Marc Isambard Brunel. But this method of working did not catch on in general manufacturing in Britain for many decades.

8. http://money.cnn.com/magazines/fortune/fortune_archive/2004/10/04/8186 795/index.htm

9. http://www.remington.com/pages/our-company/company-history.aspx

10. Roy G. Jinks and Sandra C. Krein, *Smith & Wesson* (Mount Pleasant, SC: Arcadia Publishing, 2006), p. 9.

11. http://america.aljazeera.com/articles/2013/12/23/mikhail-kalashnikovak47in ventordeadat94.html; http://www.theglobalist.com/20-facts-mikhail-kalash nikov-ak-47/

12. http://www.forbes.com/global/2003/0331/020.html

13. Philip Schreier, Senior Curator at the National Firearms Museum, interview by Barbara Eldredge, 8 November 2011, National Firearms Museum, Fairfax, VA: http://gundesigndotorg.files.wordpress.com/2013/05/barbaraeldredge_ missingthemoderngun.pdf

14. http://news.bbc.co.uk/1/hi/world/europe/6294242.stm

15. http://www.smallarmssurvey.org/weapons-and-markets/producers.html

16. http://www.reuters.com/article/2007/08/28/us-world-firearms-idUSL283489 3820070828

17. http://www.washingtonpost.com/blogs/wonkblog/wp/2012/12/19/seven-facts-about-the-u-s-gun-industry/

18. http://www.smallarmssurvey.org/fileadmin/docs/H-Research_Notes/SAS-Research-Note-43.pdf

19. James Bevan, *Rifles*, Research Note No. 38 (Geneva: Small Arms Survey, 2014).

20. Peter Batchelor and Kai Michael Kenkel, *Controlling Small Arms: Consolidation, Innovation And Relevance In Research And Policy* (London: Routledge, 2013), p. 35; it's been this way ever since President Roosevelt encouraged America to become 'the great arsenal of democracy' in Christmas 1940. The US subsequently saw a military-industrial explosion, manufacturing output doubling between 1940 and 1943, and arms production increased eight times between 1941 and 1943, reaching a level that was nearly equivalent to those of Britain, the Soviet Union and Germany combined.

21. http://www.smallarmssurvey.org/fileadmin/docs/F-Working-papers/SAS-WP14-US-Firearms-Industry.pdf

22. http://money.cnn.com/2014/07/22/news/companies/beretta-guns-move/

23. http://www.chuckhawks.com/turkish_invasion.htm

24. $11.7 billion in sales and $993 million in profits: http://www.washington post.com/blogs/wonkblog/wp/2012/12/19/seven-facts-about-the-u-s-gun-industry; http://www.ibisworld.com/industry/default.aspx?indid=662; the Bureau of Alcohol, Tobacco, Firearms and Explosives said that the number of guns made in the US rose 16 per cent between 2010 and 2011 to 6.4 million guns, and the FBI expects to run 17.8 million firearm purchase background checks in 2012 – up 9 per cent from 2011.

25. $688.3 million: http://www.ruger.com/corporate/PDF/ER-2014-02-25.pdf

26. Jeff Knox, 'Guns: A bright spot in the economy', *Firearms Coalition*, http://www.firearmscoalition.org/index.php?option=com_content&view=article&id=709:guns-a-bright-spot-in-the-economy&catid= 19:the-knox-update&Itemid=144

27. Owen Greene and Nic Marsh (eds.), *Small Arms, Crime and Conflict: Global Governance and the Threat of Armed Violence* (London: Routledge, 2012), p. 215.

28. http://www.guns.com/2014/05/02/gaston-glock-buys-wife-15-million-horse/

29. http://www.billionaire.com/guns/beretta/918/beretta-firearms-going-great-guns

30. http://www.history.co.uk/biographies/j-p-morgan

31. http://www.abc.net.au/news/2013-12-24/mikhail-kalashnikov-inventor-of-the-ak-47-dies-aged-94/5173370

32. The company Izhmash merged with Izhevsk Mechanical Plant under the new name Kalashnikov Concern; http://en.ria.ru/military_news/20140207/187298025/Kalashnikov-Concern-Estimates-Operating-Loss-of-50M-in-2013.html

33. http://www.businessweek.com/articles/2014-05-29/colts-curse-gunmakers-owners-have-led-it-to-crisis-after-crisis

34. http://www.metalstorm.com

35. http://www.nammo.com/globalassets/pdfs/product-sheets/general/nammo-brochure-2011.pdf

36. http://www.theguardian.com/uk/2009/feb/15/army-taliban-sniping

37. http://www.youtube.com/watch?v=MnA9yyE2ma8

38. http://www.lehighdefense.com/index.php/our-technology/subsonic-bullets-and-ammunition

39. http://avrockwell.com

40. http://www.nytimes.com/2011/11/27/business/how-freedom-group-became-the-gun-industrys-giant.html?pagewanted=all&_r=0

41. Ibid.

42. It then acquired DPMS/Panther Arms, Marlin, H&R, The Parker Gun, Mountain Khakis, Advanced Armament Corp, Dakota Arms, Para USA, Barnes Bullets and TAPCO; http://www.nytimes.com/2011/11/27/business/how-freedom-group-became-the-gun-industrys-giant.html?pagewanted=all

43. Cerberus has just under $160 million of equity in Freedom Group and 94 per cent ownership; http://www.bloomberg.com/news/2012-12-18/cerberus-outlay-reviewed-by-pension-after-school-massacre.html

44. http://www.freedom-group.com/2013%2010-K.pdf

45. The media people listed on the Cerberus company website – Peter Duda and John Dillard – are public-relations men.

46. http://blogs.wsj.com/deals/2007/12/18/for-cerberus-feinberg-privacy-has-a-price/

47. Tom Dias, *Making a Killing: The Business of Guns in America* (New York, The New Press, 2000), p. 11.

48. http://www.freedom-group.com/2013%2010-K.pdf

49. http://blogs.wsj.com/searealtime/2013/12/10/philippines-poverty-rate-narrows-little-despite-economic-growth/

50. http://www.vice.com/print/pr-firms-in-the-uk-are-spinning-stories-for-foreign-dictators-734

51. http://www.webershandwick.com/who-we-are/bio/eric-pehle

52. http://www.webershandwick.com/who-we-are/bio/leslie-gaines-ross

53. http://www.thefirearmblog.com/blog/2010/10/16/bushmaster-acr-recall/?utm_source=twitterfeed&utm_medium=facebook&utm_campaign=acr

54. http://www.businessweek.com/magazine/content/10_45/b4202025114499.htm; http://www.thetruthaboutguns.com/2012/06/daniel-zimmerman/remington-settles-trigger-suit/

55. http://www.bushmaster.com

56. http://www.forbes.com/sites/danielfisher/2012/12/18/bushmaster-paid-after-malvo-killings-and-may-yet-pay-again/; http://www.crimemuseum.org/crime-library/the-washington-dc-sniper

57. http://www.slate.com/blogs/the_slatest/2012/12/26/bushmaster_223_william_spengler_jr_reportedly_used_same_type_of_gun_as_adam.html; http://www.huffingtonpost.com/2012/12/25/william-spengler-had-semiautomatic-rifle_n_2362646.html

58. http://www.bostonglobe.com/business/2012/12/18/cerberus-capital-owner-steward-health-care-boston-also-owns-maker-bushmaster-rifle-used-newtown-shootings/87mKS2CJNcr2grFPAJMVTK/story.html

59. http://www.fbi.gov/about-us/cjis/ucr/crime-in-the-u.s/2011/crime-in-the-u.s.-2011/tables/table-4

60. http://www.bostonglobe.com/business/2012/12/18/cerberus-capital-owner

-steward-health-care-boston-also-owns-maker-bushmaster-rifle-used-new
town-shootings/87mKS2CJNcr2grFPAJMVTK/story.html; http://www
.huffingtonpost.com/john-rosenthal/cerberus-capital-profitin_b_4072807
.html

61. http://www.theguardian.com/world/2012/dec/19/newtown-shooting-bush
    master-gun

62. http://www.nydailynews.com/news/national/company-sell-newtown-
    gunmaker-article-1.1222849

63. http://www.ft.com/cms/s/0/7f684114-6032-11e3-b360-00144feabdc0.html?s
    iteedition=uk#axzz34npCSrgX

64. http://www.ft.com/cms/s/0/9fb20eea-6407-11e3-b70d-00144feabdc0html
    #axzz3Ehx2zdA7. And it is not just Cerberus. After Sandy Hook, Smith &
    Wesson reported sales for that year up by 43 per cent, hitting a record $588
    million. Ruger saw net sales up almost 50 per cent at $491.8 million: http://
    www.businessweek.com/articles/2013-06-26/gun-business-still-booming-as
    -anxiety-buying-continues

65. http://www.bloomberg.com/news/2013-01-09/gunmaker-tax-breaks-to-lure-
    jobs-face-renewed-scrutiny.html. Since 2007 the Group had received some
    $5.5 million in subsidies and grants.

66. http://www.cerberuscapital.com/team/stephen-a-feinberg/

67. http://www.businessweek.com/stories/2005-10-02/graphic-bio-stephen-a-dot-
    feinberg

68. http://www.nydailynews.com/news/national/company-sell-newtown-
    gunmaker-article-1.1222849

69. http://littlesis.org/relationship/view/id/203656; http://www.theguardian.com/
    commentisfree/2012/dec/20/pension-fund-investments-firearm-companies;
    http://littlesis.org/person/34771/Stephen_A_Feinberg/political

70. http://www.nrapvf.org/news-alerts/2012/09/nra-pvf-endorses-orrin-hatch-for-
    us-senate-in-utah.aspx

71. http://littlesis.org/relationship/view/id/203668; http://www.motherjones.
    com/mojo/2013/04/max-baucus-background-checks-retiring

72. http://www.bloomberg.com/bw/stories/2005-10-02/theyre-not-telling-quayle-
    jokes-at-cerberus

73. http://www.freedom-group.com/press.htm

74. http://www.sportsonesource.com/events/under40/u40winner2009.
    asp?WID=56

75. https://www.flickr.com/photos/georgekollitides/9101447597/

76. http://bfanyc.com/uploads/preview_BFA_4089_462024.jpg

77. His company certainly sells arms into Africa: http://www.remington.com
    /my-account/partners/sales/international-distributors/africa.aspx

78. http://www.conflictarm.com/wp-content/uploads/2014/09/Dispatch_IS_
    Iraq_Syria_Weapons.pdf

79. http://www.motherjones.com/politics/2013/12/freedom-group-cerberus-calstrs-bushmaster-newtown-sandy-hook

80. https://www.opensecrets.org/usearch/?q=George+Kollitides&cx=0106779074 62955562473%3Anlldkv0jvam&cof=FORID%3A11

81. http://www.businessweek.com/stories/2005-10-02/whats-bigger-than-cisco-coke-or-mcdonalds

ACKNOWLEDGEMENTS

I would like to give heartfelt thanks to my agent, Antony Topping at Greene and Heaton, and to my editor at Canongate, Katy Follain; both friends, both inspirational, both on the side of good. I am sorry for the endless morning walks and for filling your heads with dark tales of guns before your breakfasts had even been digested.

Deep thanks must go to Jenna Corderoy, my indefatigable and irreplaceable researcher on this book. A journalist with a glittering career ahead of her and a glittering mind in her head.

I'd like to thank Steve Smith and all of those at Action on Armed Violence for your help, too, in the writing of this book. Without your support none of this would have been possible.

Without the following people, this book would also not have been written. Either through your decisions as editors to send me to far flung parts of the world, your abilities as reporters to help me see the light amidst the pathways of shadows, your courage to speak out in the face of the gun's terror or just for granting me your time, insight and patience, I thank you equally and deeply:

Aage Borchgrevink, Ailsa Bathgate, Alexander Renderos, Alice Shortland, Alicia Fernández, Anne Cadwallader, Apostolos Spanos, Barbara Eldredge, Cate Buchanan, Christopher Coker, Claudia Xavier-Bonifay, David Mapstone, David Potter, David Watson,

Dorothy Parker, Elaine Potter, Flossie Baker, Frank Gardner, German Andino, Henry Dodd, Jamie Byng, Jamie Mills O'Brien, Jenny Kleeman, Jòn Pálmason, Jaz Lacey-Campbell, Lesley Levene, Mark Murray-Flutter, Michal Lee Sapir, Molly Molloy, Nic Marsh, Oren Rosenfeld, Ramita Navai, Robin Barnwell, Roy Isbister, Sam Poling, Sara Ramalho, Jon Snow, Shahida Tulaganova, Simon Reeve, Sophie Lochet, Stella Hermes, Vicki Rutherford, Will Thorne, Willard Foxton.

Those whose help I fail to mention here, the failing is all mine, but thank you to those who know.

I'd like also to thank all of those people I have not met whose work is referenced in the footnotes and pages herein. I have stood upon your shoulders. To acknowledge each and every one of you in the main body of the text or individually would take up a book in itself. I hope that I interpreted your findings accurately and that you accept that a footnote is more than that – it is a sign of my deep admiration for your journalism and research.

And then, of course, the deepest and most private gratitude must go to my family: whose patience made this possible, whose belief calmed my nerves, and whose spirits eternally lifted me. You remind me there is always something left to love.